GENESIS and the Thoughtful Christian

By: T. D. Wise

Copyright © 2015
by Professor Theophilus' Emporium of Imagination, Inc.
All Rights Reserved
Published by:
WISEWIRX MEDIA
A Division of Professor Theophilus' Emporium of
Imagination, Inc.
Magnolia, Arkansas
Cover Design by: Timothy D. Wise
Library of Congress Control Number:
2014917868

First Edition
ISBN 978-0-9908690-1-6

Lovingly dedicated to They Who Shall Not Be Named

Acknowledgements

The inspiration for this book, like all of the books I've written, came from many sources. As a young person, I was inspired by both the fiction and nonfiction work of author C.S. Lewis. Josh McDowell, the author of *Evidence that Demands a Verdict* was also an early inspiration. In discussing the relationship between religious faith and evidence, McDowell once said, "Jesus said, 'You shall know the truth,' not 'You shall ignore the truth.'" That statement has served as a kind of guiding philosophy in my own quest for a faith that is both intellectually and spiritually authentic. Dr. Herb Dickerson, a former seminary professor who served as my pastor when I was in my late teens, was a good model of that kind of faith. So was Dr. Bob Utley. Utley was a hermeneutics professor at a Christian university. He served as an interim pastor at my church for only a few months but left an indelible mark. Philip Wade, my college minister, was a mentor and encourager to me in many important ways. We have remained friends throughout the years, and I think my life would have turned out very differently if I had not met him.

I owe a debt of gratitude to students and colleagues at Southern Arkansas University for their friendship and for the ways they have both encouraged and challenged me in my ways of looking at the world. Some of them share my religious convictions, and some believe differently, but I hope all will find that I have at least been fair, honest, and kind in writing about the differences and know that I value their friendship.

The faculty at New Orleans Baptist Theological Seminary have been excellent models of intellectual engagement. Dr. Robert Stewart coordinates the Greer-Heard Forum, a series of dialogues between leading evangelical authors and leading thinkers from outside the evangelical community. The dialogue between N.T. Wright and John Dominic Crossan on the Resurrection of Christ is an example. Dr. Dennis Phelps has been especially encouraging to me in my work on this project.

One immediate source of inspiration for this book was a collegiate Bible study group I taught at my church. The students were bright, creative, and engaged. They often asked tough questions and challenged me to "up my game." Four of them formed a rock band and performed at Books-A-Million for the release party of one of the Harry Potter novels. Their group was called "They Who Shall Not Be Named." That is the story behind my rather cryptic-sounding dedication.

Dr. Dennis Hensley critiqued this manuscript in the early stages of its composition and gave me many helpful suggestions. He even had some of his colleagues review some of the chapters, and I am grateful for their feedback.

I owe my family in many ways. My mom always reads my books, whether they are any good or not. My dad is a retired chemical engineer. He could always spot plausibility problems in TV shows and movies, and sometimes that spoiled the mood a bit, but he trained me to think analytically, and that has served me well over the years. My uncle, Dan Wise, wrote his own book on faith and science a few years ago. Proofing that book for him made me think about what kind of book I might want to write if I ever tried my hand at nonfiction. My brothers, aunts, uncles, cousins, nieces, nephews, and in-laws have blessed my life in more ways than I can count.

There are many others, of course, and since this is not likely to be the last book I write, I'll save the rest of my overflowing gratitude for future volumes.

INTRODUCTION

Most of my favorite Christian authors are former skeptics who either clawed their way back to faith or fought kicking and screaming against it before finally reaching a point of surrender. These writers have been faithful friends to me in my explorations, and I suppose I envy them sometimes. One of the great fears of my young adulthood was of losing my faith and plummeting headlong into a dark pit of existential despair. My mentors, it seemed, had faced that darkness and emerged from it with a greater sense of certainty than they'd ever thought possible. I wonder sometimes if they ever really beat their doubts completely or if they just learned to manage them. Perhaps they reached a tipping point where they realized that their skepticism was no more substantial than they used to think belief in God was.

Some books about faith and science make it sound as though the evidence for God's existence is so compelling that faith is almost inevitable to anyone who looks at it honestly, and the arguments of the authors really make it look at way. On some dark nights of the soul, however, the whole faith enterprise can seem so implausible that all I can do is close my eyes and hold on out of sheer stubbornness.

In a theology textbook, I read about three models of the relationship between faith and reason: *reason then faith*, *faith alone*, and *faith seeking understanding*.[1] The assumption behind the "reason-then-faith" position is that it is almost possible to reason your way to God. Thomas Aquinas and others have attempted to formulate logical proofs for God's existence. Their approach, it seems, was based on the assumption that people make faith decisions logically and

---

1 John Newport, *Life's Ultimate Questions* (Dallas: Word Publishing, 1989), 415-416.

rationally and aren't strongly influenced by emotions or motives. In most instances, that is probably not the case. Some people's religious experiences do have a strong intellectual component, but there are usually emotional motivations as well as intellectual ones behind any religious decision.[2]

In contrast to this first group, the "faith-alone" philosophers assume reason and faith are either opposite or unrelated. As Blaise Pascal expressed it, "The heart has reasons that reason cannot know." Pascal was a strong believer in reason, but recognized that he did not arrive at faith the same way he solved math problems.[3] "Faith alone" really has reasons, of course, but they are more rooted in subjective experience than in anything resembling scientific data. This kind of faith can be passed along by sharing an experience or by telling a story, but not by objective study. Some "faith-alone" theologians do not see any value in trying to build an intellectual case for believing in God or the Bible (called apologetics) because, they say, "You can't argue someone into the kingdom." Many apologists would agree that arguing with someone is not the best way to win converts. They would add, however, that it may be possible to argue someone out of the Kingdom of God.

I once heard a story about a child who defined faith as "believing what you know isn't true." The problem with that definition, of course, is that you *can't* believe something you know isn't true. Faith can live with a certain amount of ambiguity, but there is a point at which it becomes so weighted down with unresolved contradictions that it finally collapses. That is why I would argue that it is important to feed the mind as well as the heart. That is especially true for those of us who work in academic fields where we regularly encounter evidence (or just attitudes) that conflict with Christian beliefs as we understand them. (I realize some groups of Christians do

---

2 Ibid., 415-424.

3 Ibid., 425.

not believe it is possible for a *true* Christian to lose his/her faith, but will not debate the theological issues here. Free will, predestination, and the definition of a true Christian are subjects for another discussion.)

People can get defensive when it comes to the view they hold to: "Your faith is just based on emotion, but I've actually researched mine." "Your faith is based on prideful human logic, and mine comes straight from the inerrant word of God."

The "faith-seeking-understanding approach," as advocated by theologians like Anselm and Augustine, recognizes that a human being is a unified whole with a mind as well as a heart.[4] The intellect and heart are not so separate as some would have us believe. Sometimes the heart biases the intellect, and sometimes intellectual questions must be dealt with for the heart of faith to stay healthy. There are times, however, when we find ourselves faced with questions that have no easy resolutions. Even leaders we respect are sometimes divided over the answers.

The issues surrounding the book of Genesis are among the most vexing for many Christians. If the sun was not created until the forth day of creation, why was there light on the first? Where did Cain get his wife? Did ancient people really live over nine hundred years? If the internal questions were not enough, such scientific discoveries as dinosaur and hominid bones and such archeological discoveries as the tablets containing the Epic of Gilgamesh raise thornier questions still. (Gilgamesh is older than the Hebrew language but has a serpent and fruit story and an ark story that resemble the ones found in the biblical narratives.) In the end, nobody really has the final word on how to resolve these issues. Denominational background, personality type, and education level all play a role in how we react to these problems and whose solutions we find the most credible, but no approach is without its problems.

I will hasten to add, however, that readers should not be discouraged by the lack of consensus on the secondary issues.

---

[4] Ibid., 428-429.

Arguments for the role of God in the creation of the universe, the emergence of life, and in the origins of human consciousness are easier to defend than, say, a six-day creation event or a planetary flood and there is generally more agreement on those, at least among Christian scholars. Many apologists like Lee Strobel, in *The Case for the Creator*, have chosen to concentrate on these big-picture questions and avoid the areas where simple answers are not as readily available.

Other Christian apologists have chosen to avoid the Old Testament altogether and concentrate on the life and resurrection of Jesus. "If Jesus really did rise from the dead, and you can build a convincing case for that," they would say, "what does it matter about dinosaur bones and Gilgamesh tablets?" That is certainly a good point. It is also worth noting that writing style used by the New Testament authors is probably more similar to what Westerners are familiar with. The Greeks wrote in a precise and concrete way that was copied by both the New Testament authors and the European educational systems of the Middle Ages. Americans and Europeans, therefore, think more like New Testament Greeks and less like Old Testament Hebrews.

I struggled for a while about whether to write this book. It would certainly have been easier to have found a safer, less controversial subject. I chose to tackle the harder questions because I encountered them among university students and faculty and felt that they deserved honest answers even if they were not always conclusive ones. Though most of the pastors and teachers in the churches I grew up in welcomed honest questions, I have found that that is not always the case anymore. In the war of words between liberals and conservatives, evangelical intellectuals have become a high risk group in many churches. Some church leaders might think, "Let them leave. They never fit in here anyway." Given the influence academics have on society as a whole, this attitude is short-sighted. What's more, there is no point in it. Though popular culture presents faith and reason as enemies, the church has had a rich intellectual heritage that goes back for centuries. Galileo and the Scopes Trial are held up as examples of "how it always is," but the real relationship between the two is much more complex.[5]

The answers I give or avoid giving will not satisfy some people. In some instances, I have refused to align myself with positions other Christians I know hold to because the evidence for those positions did not seem convincing to me. In other cases, I have refused to take a position at all because I couldn't really build a solid case one way or another. I have, at least, tried to propose reasonable possibilities.

In the final analysis, it does not matter that we agree on everything. This book, for me, represents a point in a journey, and I'll no doubt revise some of what I've written as further evidence comes to light or as I dig more deeply into some of the topics I've tried to cover. If I make you think and help you develop your own ideas further, I've done my job as a writer and teacher. If, at any time, you find that reading about the issues I've discussed here is harming you more than helping, lay the book aside. There are questions that would have disturbed me deeply twenty years ago that do not bother me so much now. This book may be what you need at this point in your life, or it may not be. If you're already wrestling with some of the issues discussed here, I hope I can give you some direction. At the very least, I can give you the assurance that you're not alone in asking your questions or weird for asking them. All in all, I hope most readers will find it to be an interesting and educational journey into a mysterious lost world.

---

[5] See, for example, Ronald L. Numbers, *Galileo Goes to Jail and Other Myths About Science and Religion.* Cambridge, MA: Harvard University Press, 2010, and Timothy Larsen, *Crisis of Doubt: Honest Faith in Nineteenth-Century England.* New York: Oxford University Press, 2009.

CHAPTER 1:
Apples, Oranges, and the Pursuit of Truth

According to a 2012 Gallup Poll, 47% percent of Americans believe in "evolution" and 46% do not. Of those who said they believed in evolution, 32% still believed that God had some kind of a role in the process of creation whereas 15% (presumably atheists, agnostics, and followers of non-theistic religions) did not believe God had been involved at all. Naturally, these percentages vary with both education level and political party.[6]

As strongly as people seem to feel about evolution, perhaps we had better stop and define the term. Just what do we mean when we use the word evolution, anyway? What does it mean when someone says he or she *believes in* evolution? For many Christian groups, evolution is a politically charged buzzword like *communist* and is often equated with atheism. It represents the war between people of faith and intellectual skeptics. It means we "came from monkeys" instead of being created by God. *Abiogenesis*, the chance emergence of life from nonliving chemicals (the primordial soup, etc.), is usually lumped in with evolution, but technically they're two different things. Many of the arguments we hear against evolution are really arguments against abiogenesis.

As a technical term, the word evolution has a variety of meanings. It can refer to common ancestry with minor variations of similar animals found in various parts of the world (e.g. horses, donkeys, and zebras) or it can refer to a comprehensive theory of origins that explains all of life. Many

---

[6] Frank Newport, "In U.S. 46% Hold Creationist View of Origins," Gallup Organization, entry posted June 1, 2012, http://www.gallup.com/poll/ 155003/Hold-Creationist-View-Human-Origins.aspx [accessed July 9, 2012].

Christians would argue that microevolution, the minor variations within or between species (like the horses, donkeys, and zebras in the example), is a proven fact, but do not accept macroevolution, the development of radically different species from the same ancestors, or Darwin's General Theory of Evolution, a theory that explains all life on earth as sharing common ancestry.[7]

That is not to say that there are not Christians who accept Darwin's General Theory, but many of these would make a further distinction between evolution as a scientific theory and evolution as a philosophy of life (called *evolutionism*). Evolution, they say, explains the ways living organisms adapt over time, but evolutionism goes beyond biology. It is a comprehensive worldview that makes philosophical assertions about the nature of reality and the purpose and destiny of humanity: Humanity is an accident of genetics. There is no ultimate purpose to life. There is no room in nature for the kind of God described in the Bible.[8] These are not statements that can be proven in a lab, but interpretations some scientists make based on the philosophical lenses through which they view the world. This is an important distinction, and we will return to it in the pages that follow.

Since the time of the Scopes Trial, the issue of origins in education has come up repeatedly in legislative bodies around the United States. Bills calling for equal time for the teaching of creationism have been submitted to the legislatures of over thirty states, states as culturally diverse as California and my own beloved Arkansas.[9] (I realize both states are stereotyped and not always fairly, but it does make for an interesting study in contrasts.) Homeschooling of children has become

---

[7] John Newport, *Life's Ultimate Questions* (Dallas: Word Publishing, 1989), 138-139.

[8] Ibid., 137.

[9] Ibid., 122.

increasingly popular among evangelical Christian families in the U.S., and, according to a *USA Today* article written in 2010, the majority of home school families use some form of creationist literature to teach their children science.[10] (The term creationist is used in a general sense here. Some writers use it only to refer specifically to those who adhere to a strict young earth/flood geology model, whereas others use the term broadly to refer to any approach to science that allows a place for God in the creative process.)

Though some observers might wonder what all the fuss is about, the reasons for alarm are easy enough to understand. Surveys conducted between the 1960s and the 1980s showed a rise in secularists (those who claim no religious allegiance) from 2% of the American population in 1962 to 11% by 1990.[11] A later survey, taken in 2009, showed that the number had risen to 16%.[12]

A study by Robert Wuthnow linked the rise in secularism to an increase in the number of Americans who received university educations.[13] Though more recent studies have shown that university education no longer has any significant correlation to loss of religious faith[14], the trend was prevalent for a number of years and has had an impact on American

---

[10] Dylan Lovan, "Top Home-School Texts Dismiss Evolution for Creationism." *USA Today.com.* Posted 3/8/2010, 2:30 p.m. http://www.usatoday.com/ news/religion/2010-03-08-home-school-christian_N.htm [Accessed July 9, 2012].

[11] Robert Wuthnow, *The Restructuring of American Religion* (Princeton, NJ: Princeton University Press, 1988), 155. Cited in James Davison Hunter. *Culture Wars.* (New York: HarperCollins, 1991), 76.

[12] Philip Yancey, *What Good is God?* (New York: Faithwords, 2010), 3.

[13] Wuthnow, 76.

[14] Christian Smith, *Souls in Transition* (Oxford: Oxford University Press, 2009), 248.

culture. It is no wonder many Christians see education as an important key to the survival of Christianity as a cultural force. Science education, in particular, has been seen as a target for reform efforts.

On a parenthetical note, not everyone who checks "no religion" on a survey has necessarily abandoned belief in a Supreme Being or, at least, in a spiritual reality. Many have, instead, rejected institutional religions in favor of a cafeteria approach to spirituality. It is not uncommon to hear such people describe themselves as "not religious, but spiritual." Their rejection of religion has less to do with science than with their personalized approach to spiritual authority, the idea that "My spiritual life is my own business, and nobody can tell me what to believe."

I generally encounter the "build your own religion" approach more often among my liberal arts friends than among business students and faculty, and sympathize with it up to a point. I think most people do. Given the choice between forced conformity (Accept this doctrine or die!) or total anarchy (Believe anything you want.), most people would probably opt for the second, but they would qualify the "anything" to varying degrees. Some would also be quick to add that the freedom to believe anything does not mean that all beliefs are equally credible. "You're welcome to believe your parents were really space aliens if you want to," they might say, "but don't expect me to."

Like New Agers and postmoderns, evangelical Christians place a great deal of emphasis on "heart" experiences they believe are personal encounters with God. "Christianity," they say, "is not a religion but a relationship." Some people outside of evangelical circles probably see this as a boast, but evangelicals do not see it that way. For them, it is more like the "not religious but spiritual" claim of postmoderns. It is a rejection of externally-imposed ritual in favor of personal experience.

Evangelicals do not *really* reject religion, however. They depend on the Bible and church teachings to set boundaries on their spiritual experiences. Without some commonly recognized source of spiritual authority, they would say, most of us end up worshipping our own reflections. Even with that kind of authority, unconsciously remaking God in our own images (God is a Republican. God is black. God is a woman.) is still a danger, and not even Christian leaders are immune. Sometimes, in fact, the prestige that comes with their positions makes them even more vulnerable. I intend to explore the authority/autonomy question further in a later book, but, for now, I'll restrict my focus to the faith/science questions surrounding the book of Genesis.

Returning to that subject, many Christian leaders have attempted to "bring the Bible back into science," but what many of them do not consider are the practical challenges inherent in forcing a shotgun wedding between science and religion. Scientific and religious thought, language, and methods differ in important ways, and that must be clearly understood (to borrow from Dickens) before anything wonderful can come from the journey we are about to undertake. Christians in scientific fields encounter these differences every day but often find little sympathy from Christians who are not involved in the sciences.

These first two chapters will address the "apple and orange" differences between scientific and religious thought that Christians face anytime they try to integrate the two. ("Comparing apples and oranges" is a popular southern metaphor for comparing two different categories of things.) Even if homeschool parents or Christians in private Christian institutions have the freedom to incorporate religion into discussions of science, how should they go about it? If a person of faith is also a scientist, what issues should she consider in resolving the apparent conflicts between the ways Christians and scientists approach the world? Much of this chapter, as well as the next, will be dedicated to the discussion of seven challenges inherent in the marriage of faith and science and to possible ways to address them. Though they do not eliminate the inherent tension between the two, they can at least provide structure and coherence to the conversation which are often lacking in public debates. These, then, are the challenges:

> 1: "Galileo-type" science is based on observation, but the Christian God is invisible.
>
> 2: Modern scientific research takes place in mostly secular environments.
>
> 3: Christians seek miracles, but scientists want to understand how things work.
>
> 4: Scientists believe reason is reliable, and some Christians do not, but we're all stuck with it regardless.
>
> 5: Scientists believe reality is realistic, but not all Christians agree.
>
> 6: Scientific language is precise, but biblical language is nuanced.
>
> 7: Science is dispassionate about results (ideally, at least), but religion demands dedication to doctrine.

Now we'll start with the first:

Challenge #1: "Galileo-type" science is based on observation, but the Christian God is invisible.

What did Galileo and the scientists of his day do that was so important that we're still dealing with the results today? The short answer is that they changed the way science was done. Before Galileo's time, science was mixed with (some would say "mired in") theology and philosophy, but Galileo and the astronomers of his era changed the rules of the game by basing their conclusions on carefully recorded observations of heavenly bodies and on mathematical calculations. The idea was to reduce the likelihood of human error by restricting science, as much as humanly possible, to observable phenomena. Anything that could not be detected, or at least inferred, by observation and calculation lay outside the realm of their new science. When we talk about the difference between an art and a science, that is what we mean. Science moved out of the philosophy department, we might say, and moved in with math and engineering.

This shift placed God outside of the boundaries of hard science because God, as described in the Bible, is invisible to human perceptions unless he wants to be seen, and he does not generally perform on command. He goes out of his way to reveal himself to some people yet, for his own purposes, conceals himself from others. As annoying as it is to some of us, faith is still an essential element of religion. That doesn't mean religious questions don't matter, just that observation-based science isn't equipped to answer those questions.

When Galileo's observation-based science turned out to be wildly successful, and a series of new discoveries reshaped humanity's understanding of the universe, it was all too easy for people to assume that science would eventually be able to explain everything without reference to a God. If they could explain weather, planetary rotation, and maybe even the origin of life without falling back on God as an explanation, (some concluded) God was out of a job. Darwin's theory of evolution, in the words of Richard Dawkins, made it possible to be an "intellectually fulfilled atheist."[15] Is that really true?

The technological advances of the past five centuries have taken us from horses and buggies to the moon, but we should not let our success make us overconfident. Astronomer Simon Newcomb once wrote that pretty much everything of importance to astronomers had been seen and measured, and all that was left was to combine the knowledge. Perhaps astronomers would find a few more comets, but the overall picture of the universe was settled. The statement does not really sound all that outrageous until we find out Newcomb wrote it in 1888 before Hubble discovered our galaxy is one among millions.[16] As Newcomb's declaration shows us, the

---

[15] Richard Dawkins, *The Blind Watchmaker* (New York: Norton, 1986), 6.

GENESIS and the Thoughtful Christian/19

idea that modern humans have a pretty good grasp on how the universe works can be a dangerous conceit, but there are still many mysteries left to humble us.

Though Sir Isaac Newton explained the effects of gravity centuries ago, scientists have never found a way to detect gravity itself nor do they completely understand its properties.[17] Some scientists now believe that 95% of the universe is composed of dark matter and dark energy. Like gravity, these "dark" components are completely undetectable. They are theorized to exist because of measurable behaviors in visible matter and energy that cannot be explained by anything that has, so far, been detected.[18] Science apparently has plenty of room for invisible forces, but *personal* invisible forces (assuming they exist) are not something science is equipped to deal with. Impersonal forces are repetitive and at least somewhat predictable when they are understood. Scientific observation relies on that type of predictability. Personal beings are driven my motives and desires and do not fit into the models and formulae that characterize the scientific method. Scientists like Galileo, Kepler, and Newton believed in letting science be science and letting theology be theology. (Render unto science the things that are science's, and unto God the things that are God's?) Maintaining those boundaries, or even deciding where they lie, has not been easy. Scientific discoveries can have implications science alone is not equipped to address.

Before I move on to the next point, we should note the difference between "hard" sciences and "soft" sciences. Hard sciences like physics deal with observable phenomena. Soft

---

[16] Simon Newcomb. "The Place of Astronomy Among the Sciences," *The Sidereal Messenger*, 7 (1888): 69-70.

[17] Charles E. Hummel, *The Galileo Connection* (Downer's Grove, IL: Intervarsity Press, 1986), 142, 154.

[18] Adam Hart-Davis, *Science: The Definitive Visual Guide* (London: Dorling Kindersley, 2009), 398.

sciences like psychology deal with intangibles like feelings and perceptions. Because of this, many hard scientists do not consider the soft disciplines to be sciences in the proper sense, but we will not argue semantics here. Disciplines like history, archaeology, and crime scene investigation operate on both sides of the boundary by using scientific instruments and methods to study areas of life—including religious experiences—that are not open to direct observation. When scientists say religion lies outside the realm of science, they really mean it lies outside the realm of *hard* science. A psychologist's study of near-death experiences and an archaeologist's study of sites like Jericho may touch on religious questions but cannot be said to prove or disprove the existence of God in any scientific sense. Naturally, the way one defines the word "prove" differs from one discipline to another. Proving that Ivory soap floats (to borrow from Josh McDowell) is different from proving that someone committed a crime. Lawyers use expressions like "prove beyond a reasonable doubt." This use of the word proof allows for more subjectivity than would be the case in a lab experiment, and most events involving the chosen actions of living beings would fall into this category.

Some Christian readers might consider my definition of science to be too restrictive. The expression *methodological naturalism* has often been used to describe the "observation and calculation" approach I have described here. Methodological naturalism is when scientists limit their investigations to natural causes and supernatural causes are excluded from the discussion at the outset. Though some scientists eliminate the supernatural because they do not believe anything beyond nature exists, others stick to natural causes because they are readily observable. "I'm not saying the supernatural couldn't exist," they would say, "just that science isn't equipped to make that kind of determination. You can't put God under a microscope." Some might add that science is clean. Why muddy it up with religious arguments? Just stick to the evidence. I respect that perspective and understand the practical reasons for it. The problem is when scientists start out by limiting their investigations to natural cause and end by claiming that they have disproved the existence of the supernatural when it was never on the table to start with. That line of reasonsing is a bit like writing a book about great artists of the world, limiting yourself to American artists, and then claiming that all of the world's great artists were American.

If you are a Christian, a Muslim, or a follower of another supernatural belief system, your definition of science might not include methodological naturalism as a ground rule, and that's fine. My point is that before we say something is "scientific" or "unscientific," let us at least be clear on just what we mean. Most of us take it to mean something like, "Hard evidence supports this," not, "In the opinion of a scientist."

Some Christian readers might object to the idea of God being invisible to science. Doesn't the book of Romans say that the invisible attributes of God are present for all to see in the world of nature and that people have no excuse for not believing in him? That is a point worth discussing. Some people take the statement in Romans to mean that it is possible to prove God's existence by observing nature. Others of us would say the world of nature makes the existence of a Supreme Being universally accessible to human intuition, but does not offer proof in the scientific sense. That would make the existence of God evident to the young and old, and to the educated and the uneducated alike, but would still leave room for disbelief. We could argue over the strength of the evidence found in nature or the extent to which someone would have to willfully ignore that evidence to remain an unbeliever, and many have. One thing Christians and skeptics share is the inability to ever completely step outside of our own biases. As biased as we all are, however, we still manage to change our minds from time to time.

Let's sum up this discussion before we move on to the next challenge: If methodological naturalism places God outside the domain of scientific research, does that mean scientific methodology cannot be used at all in discussing religious questions? Not necessarily. As I mentioned earlier, there are soft sciences and hybrid disciplines that combine scientific measurement with subjective attempts to reconstruct past events. The investigation of biblical events, including miracles, would fall somewhere in this realm. Some people, because of their preconceptions, would find all miracle stories equally outlandish. Others, however, would at least give consideration to miraculous explanations for some events, especially when no plausible natural explanations could be found or when the natural explanations sound even more outlandish than the supernatural ones. If the choice came down to God, aliens, or objects popping in or out of existence from the quantum ether for no apparent reason, for example, which would you choose?

Challenge #2: Modern scientific research takes place in mostly secular environments.

The practical problems inherent in measuring God are only one consideration. The venues in which scientific investigation takes place are another. In Galileo's day, Catholicism was the law of the land, but modern universities and academic societies are no longer confined to the predominantly Christian and Roman Catholic world of pre-Reformation Europe. In Galileo's time, scientists only had to worry about accommodating one form of Christianity, but modern scientists may be Roman Catholic, liberal or conservative Protestant, Mormon, Jewish, Muslim, Hindu, Buddhist, Confucian, atheist, or agnostic. Science, which is mostly conducted in secular universities and labs, is a cross-cultural, interfaith endeavor. Even if it were not for the methodological issues we just discussed, leaving religion out of the discussion is simpler for strictly practical reasons.

Groups of Christians that make organized efforts to "bring God back into science" are usually doomed at the outset because of the secular/multifaith environments in which science is conducted. There is, admittedly, a double standard among some scientists who feel that it is acceptable to bash religion but not to endorse it, and this can be frustrating for people of faith. The matter of what is permissible and what is taboo in the realm of science is not limited to religious issues, however. Speculation about extraterrestrial life, alternate universes, and extrasensory perception all operate in a zone between respectable science and quackery. Biologists who spend too much time delving into Bigfoot studies are likely to face the same kind of scorn some Christian scientists have reported. The Intelligent Design movement is an interesting case study in the border war between science and religion, as well as in the actual and perceived differences between scholarly investigation and pop science.

It is worth noting, however, that some ideas, like the Big Bang theory, do manage to catch on even if they were originally proposed by Christians. One of the first scientists to suggest the "cosmic egg" idea in 1931 was a Belgian priest named Georges Lematre.[19]

Because of the underlying differences in belief, some Christians believe that Christians in the sciences should withdraw from the mainstream altogether and form their own Bible-based scientific institutions. There are Muslims who feel the same way about "godless" western science. Some of them have developed Islamic science institutions. Withdrawing from the mainstream of science does not solve all of the problems, however. Even if Christians start with the same core beliefs, they do not all agree on how God operates in the world of nature.

Young earth creationists believe God created the universe 6,000 to 10,000 years ago and that humans coexisted with dinosaurs. Progressive creationists believe the earth is billions of years old and that God intervened at various points to create new animal species. As one science teacher I know told a colleague, "What you call punctuated equilibrium, I call progressive creation." Theistic evolutionists believe God set the universe up in such a way that it had everything it needed

---

[19] Martin Rees, *Universe: The Definitive Visual Guide* (New York: DK Publishing, 2008), 96.

to produce life. Some of them use the expression "fully-gifted creation" to describe their philosophy. God's role, to their way of thinking, is so seamless that it seems invisible so far as creation is concerned. They do not necessarily believe, however, that God does not intervene, miraculously or otherwise, in the personal lives of individuals, or that the universe is not governed by an overall plan. They simply believe God took a more subtle approach when it came to creation, choosing to carry out his plans through the workings of natural systems he created rather than overriding them. God makes a difference to them in their personal lives, but not in ways science can detect.

Having taught marketing classes, I could probably write an entire book discussing the advantages and disadvantages of having Christian alternatives to everything from universities to breath mints. Some Christians want to isolate themselves from worldly influences by surrounding themselves with Christian versions of just about everything. They can work in Christian companies, send their children to Christian schools, listen to Christian radio stations, read Christian suspense thrillers, exercise in church gymnasiums to Christian workout videos, and so on. Some readers are probably asking "What's wrong with that?" while others sigh and shake their heads. I admit that there are some areas of Christian consumerism (e.g. Christian novels) that I'm quite comfortable with, but others make me uneasy. The main advantage of Christian branding, of course, is that it is tailored to the needs (or wants, at least) of its target market. The main risks are isolation from mainstream culture (which limits one's ability to influence mainstream culture) and the lowering of quality standards that commonly results from competing in small and isolated markets.

The idea of "branded truth," packaging or slanting facts in a way that appeals to the religious and political views of certain groups, is often taken to unhealthy extremes, and I'm sorry to say that some Christian groups are as guilty of this as anyone else. Kurt Wise is a young earth creationist who attended Harvard and studied geology under Stephen Jay Gould, interestingly enough. (Wise and I are not related as far as I know.) Wise was dismayed when he found one of the leading young earth creationist groups making unjustified claims in its literature. At first, he said, he thought it was unintentional. When he sought them out and confronted them about it, he discovered that they were doing it deliberately. After a long conversation with one of the speakers, the man admitted he was presenting misleading information. Then, a week later, Wise heard the speaker using the same arguments again. To this creationist speaker, it didn't matter if the facts he used were true. It was all about getting people to what he considered to be the right conclusions. Science, for him, was just stage dressing.[20]

Three pieces of evidence that often appear in creationist presentations are a petrified riverbed near Glen Rose, Texas, that supposedly contains both human and dinosaur footprints, a photograph of a Japanese fishing boat hoisting up the body of a dead dinosaur, and a piece of a Tyrannosaurus skeleton that contains blood cells and blood vessels. Todd Wood, himself a young earth creationist, wrote an article in the Creation Research Institute's quarterly debunking the dinosaur footprints and the dead dinosaur. The footprints in Texas are too badly smeared to be identified as human, and most scientists think they're the prints of a smaller dinosaur. Wood located the technical report on the Japanese "discovery" and the supposed dead dinosaur was really a badly decomposed shark. Wood was shocked when the other members of the institute wrote letters, attacking his report.[21]

The story about the discovery of Tyrannosaurus bones with blood vessels, collagen, and blood cells intact is actually true, and Dr. Mary Schweitzer, the scientist who made the

---

[20] Tim Stafford. *The Adam Quest* (Nashville: Nelson Books, 2013), 15, 16.

[21] Ibid., 38.

discovery, is a Christian. Schweitzer says, however, that the age of the fossil was never in doubt. Multiple lines of evidence, she says, show the bones are ancient and their discovery does not prove that dinosaurs roamed the earth only a few thousand years ago. Though she is used to other Christians not understanding her work, she walked out of a church once when she heard a pastor preaching about how her discovery had shown, once again, just how stupid scientists are.[22]

It is hard to understand why some speakers spend so much time trying to prove that humans and dinosaurs coexisted until fairly recent times. If some dinosaurs had managed to survive until modern times, it would not prove anything one way or the other about the age of the planet or the accuracy of scientific dating methods. Crocodiles have supposedly been on earth since the dinosaur age, and they are still here. If a scientist were to discover good evidence that a few dinosaurs had managed to survive until relatively modern times, he or she would be an instant celebrity. The idea of a cover-up is not plausible.

As I was discussing these issues with a physicist friend, he gave me a more chilling example of "science for a cause." He said the Nazis during World War II developed their own science to support the Nazi cause. Nazi science, they said, was different from Jewish science. As far as the facts and data go, my friend insisted, science is value neutral. Anyone who performs the same experiment using proper scientific procedures will get the same results no matter who he or she is. That, to him, is the value of science done right.

As long as scientists confine their discussions to the evidence itself, science remains philosophically neutral, but when they begin to answer the "So what?" and "What does it mean?" questions for the rest of society, philosophical slanting is probably unavoidable. According to my friend, a scientist who respects the boundaries of his profession will avoid making sensational claims like, "We have disproven the

---

[22] Ibid., 109-114.

existence of God." Unfortunately sensationalism sells. Richard Dawkins' "scientific" attacks on religion have topped best seller lists. Even in academic circles, outrageous theories can generate more interest than mundane ones and sometimes have an easier time getting funding, but at least there are some boundaries against completely unfounded claims that do not exist in popular media. If a scientist makes an outrageous claim, he or she better have the evidence to back it up before a community of scientists who hold competing views. That is the strength of the academic review process.

With the dangers of "branded truth" in mind, readers might ask, do I still think we need Christian organizations that wrestle with scientific issues? Actually, I do because Christians have philosophical questions that are unique to Christians, and they need people who are knowledgable about both science and Christian theology to address those questions. I would say, however, that people in these organizations need to do a better job of educating their audiences about what science is, how it works, and about what it can and cannot do. They should avoid sensationalism and bandwagon approaches in favor of sober discussion and honest analysis. Some knowledge of Hebrew literature and culture would also be advantageous, but many of the debates leave literary considerations out altogether. On the whole, I think Christian universities and seminaries generally do a better job of educating audiences than many of the groups that are dedicated purely to producing creationist materials. Though we take differing positions on other matters, I think Kurt Wise and Todd Wood would agree with me on this.

Challenge #3: Christians seek miracles, but scientists want to understand how things work.

Christians, as people of faith, look for the hand of God in all areas of life. We sometimes use the word "miracle" to describe something as natural as a beautiful sunset. Scientists, however, study the universe more as someone would study a machine. Many scientists were the type of children who took their toys apart to see how they worked. Curiosity, for them, is a driving passion.

Once, I was discussing a paper in front of a graduate-level economics class when the teacher stumped me with a question. I jokingly told him, "God made it that way," and the class laughed. My professor, a devout Muslim, smiled. I still wonder what he was thinking. Some scientists might not be so amused. If you tell a scientist, "It's a miracle, and you'll never understand it," you might as well tell him, "Give up and go home." Explanations such as "It's a miracle," or "God did it," for many scientists, are just pious-sounding excuses for intellectual laziness.

A scientist who is also a Christian would simply smile and say, "Yes, but *how* did God do it?" He or she would continue to search for the kind of answer a scientist's tools can measure. Such "natural miracles" as genetics and weather are a scientist's stock in trade. The spontaneous acts of an omnipotent God, a God with plans and a personality, are not. If explaining the universe as operating purely by chance puts God out of a job, explaining it by falling back on miracles puts scientists out of a job. Viewed in that light, the turf war between science and religion is somewhat inevitable as far as methods and mindset go. It comes down to a conflict between the drive to know and the insistence that we never can. Curiosity collides with the reverential preservation of mystery.

In a later chapter, we'll note the reactions of scientists to the Big Bang Theory and the later discoveries that helped to support it. Some Christians do not like the Big Bang Theory, and it would surprise them to discover that some scientists do not like it, either. As astronomer Robert Jastrow points out in *God and the Astronomers*, some scientists see it as hitting a wall.[23]

If, as Christians, we believe God is capable of working through natural laws or intervening miraculously, how can we

---

[23] Robert Jastrow, *God and the Astronomers* (New York: W. W. Norton, 1992), 107.

GENESIS and the Thoughtful Christian/29

determine which way he was operating in any given instance? There is no clear answer to this question. It is still useful to recognize the difference between the "fact-seeking" and "miracle-seeking" mindsets, the motives behind them, the tension between them, and the need for both.

Why, some might ask, do we need the fact-seeking mindset at all? Isn't it better to view all of life as a miracle and to give God glory for everything? That is a good question, and there are good answers. The fact-seeking scientific approach has allowed us to develop modern medical technology and modern methods of treating mental illness. Previous generations would have attributed both physical and mental illness to the activity of evil spirits. Although Christians still allow for the possibility of demon possession, it is not the first thing doctors look for when they are treating their patients. I am, I admit, thankful for that. I am also thankful that we no longer try people for witchcraft.

Is it bad that we no longer see the natural world as a magical place? The Northern Lights are so breathtakingly beautiful that previous generations might have been driven to worship them. Hodiak, I believe, was supposed to be the goddess of the Northern Lights. To us, they are just another phenomenon of nature. We must keep in mind, however, that part of the purpose behind the book of Genesis was to get people to stop worshipping such natural phenomena as the sun, moon, and oceans, and to worship the invisible God beyond them. In that way, Genesis laid the foundation for the scientific mindset by making nature a creation of God and not a pantheon of gods.

Another point is worth bringing up here. When the Bible tells of the acts of God, it includes both natural acts and miraculous signs. For simplicity's sake, I will call then natural miracles and sign miracles though I realize this is something of an abuse of the word miracle.

Natural miracles are events attributed to the providence of God that do not actually involve any suspension of the laws of nature as we understand them. Most of us would not consider them miracles in the sense that we usually use the word. Psalm

104 says God causes the rain to fall and the plants to grow. Science offers natural explanations for both rainfall and plant growth, but few of us would argue that science has disproven this Psalm. We would say, instead, that the Psalm is speaking somewhat poetically about God as the ultimate cause of everything, whereas science deals with the immediate mechanics.

Sign miracles are events that interrupt the routine operation of nature to reveal the presence of a powerful Personality beyond nature. These are the wondrous signs we associate with the ministries of Jesus and Moses. Attempts to explain these miracles scientifically miss the point of what a sign miracle is meant to be.

Which category, then, do the creation miracles in Genesis fall into? Those who favor a more literal interpretation of the book would place them all into the same category as the sign miracles of Jesus and Moses. Nothing in nature, according to their perspective, could have arisen on its own, without God's miraculous intervention, so the existence of nature itself is a sign from God in the same sense that the resurrection of Christ was.

Theistic evolutionists (people who believe in God and evolution) view Genesis as mostly metaphorical, placing many of God's creative acts into the same "natural miracle" category as rainfall and plant growth. Evolution, they would say, is a process that God ordained to accomplish his providential purpose. Science, they would say, reveals the mechanisms, and Genesis reveals the purpose. Progressive creationists would split the difference with God initiating major changes miraculously into the operation of nature and then allowing nature to operate on its own for a while before intervening again.

None of these views rules out sign miracles in events like the Exodus or the Resurrection. What many ask, instead, is whether a sudden miraculous creation or a gradual, providential creation is a better fit for the evidence surrounding a particular creation event. If the universe was created all at once in the recent past, some would ask, why are there so many artifacts that point to an older universe?

Challenge #4: Scientists believe reason is reliable, and some Christians do not, but we're all stuck with it regardless.

One of the ground rule prerequisites for science is that human reason is, in the main, trustworthy. Some Christians would disagree with this. Human reason, they would say, was so marred and distorted by Adam's Fall that it is completely unreliable, and the Bible is the only thing we can trust. This idea does have biblical roots (See Ephesians 4:17-19, 1 Corinthians 2:6-3:23, Colossians 2:8, and 2 Thessalonians 2:11), but the currently popular version of this teaching has its roots in the work of Cornelius Van Til. Van Til taught when it came to reason, there was no neutral common ground between Christians and non-Christians.[24] The Christian mind and the depraved mind of the non-Christian were so different, from his point of view, that there was no point in Christians and non-Christians trying to have any kind of reason-based discussion. (Note that Paul did have such a discussion with the philosophers at Mars Hill in Acts 17:16-33.) Gordon Haddon Clark, another influential teacher, took the rather extreme position that the only things anyone could truly know were the ones found in the Bible.[25]

This view of human reason (the street version of it, at least) is widespread among certain groups of Reformed (neo-Calvinist) Christians, and is used by some creationist groups to justify force-fitting or even misrepresenting scientific evidence to fit a literal reading of the Bible. "Is it better to be loyal to manmade science," they might ask, "or to be faithful to the word of God?" Perhaps we should ask if it is better to slant the truth in a way that supports the Bible or to tell the whole truth,

---

[24] Kenneth D. Boa and Robert M. Bowman. *Faith Has Its Reasons* (Grand Rapids: InterVarsity Press, 2001), 240-243.

[25] Ibid., 245.

even when it seems to challenge the Bible, and have enough faith to believe that God's truth will win out in the end.

As an educator, I confess I would rather see someone avoid science entirely than see them using it incorrectly or unethically, even for a good cause. Some Christians who work in scientific fields choose to compartmentalize the two areas of their lives and "let science be science, and let the Bible be the Bible." This, for many of us, is preferable to seeing them force-fit the two together in bizarre ways that make mockeries of both the Bible and science. Some people choose to put off resolving the problems for now and hope future discoveries will make it easier to reconcile the apparent contradictions. Others formulate possible ways to reconcile the conflicts but avoid presenting their proposed solutions as the final word for Christians or the only solution that respects the Bible as God's word. That's what I'll be doing later as I discuss issues that do not lend themselves to easy resolution.

There is another point I would like to make before we move on: Though I do not believe the Bible's teachings about the flawed nature of human wisdom should be used to justify the misrepresentation of evidence no matter how noble the goal, I cannot deny that human beings are flawed. Our biases and personal motives (call it sin, fallen nature, or whatever you prefer) do compromise the way we use our cognitive abilities. Becoming a Christian and turning to the Bible may help with some kinds of problems, but it is naïve to think a conversion experience solves all of the problems.

Anytime Christians go beyond *what the Bible says* to *what the Bible means*, some method of interpretation is unavoidable. That means they have to use their biased and imperfect brains to understand it. They may try to get around their biases by relying on careful scholarship or they may depend on the Holy Spirit to illuminate the Scriptures for them and rely on (hopefully) divinely-directed intuition. These methods are not infallible. Our knowledge is incomplete and the insights we think come from the Holy Spirit are easily slanted by our own feelings and motives. Human reason is flawed, but we are stuck with it because one cannot think without a brain. Some Christians think they can get around this by taking every verse in the Bible literally. "I don't interpret the Bible," they said, "I just read it and believe it." That is a conceit. The choice to read a passage literally *is* an interpretation. Furthermore, nobody I know really takes *everything* literally. In Matthew 5:29-30, for example, Jesus told his listeners to cut off their hands and gouge out their eyes if they caused them to sin. Few Christians I know would take this statement literally. Roman Catholics believe the bread and wine eaten in the communion meal literally become the body and blood of Christ in some miraculous fashion, but most evangelicals would not even think of interpreting the account that way. The order of the creation events in the creation week narrative differs from the order given in the story of Adam's creation. So do some of the other details. Are we supposed to take the order given in the seven-day creation story literally and the order given in Adam's story more fluidly?

When asked if I believe in a "literal Bible," my answer is that I try to take the literal parts literally and the figurative parts figuratively. I try to use context and common sense to tell the difference, but sometimes what looks like common sense to me does not look that way to someone else. In spite of our best efforts and intentions, our incomplete knowledge and biases keep us from ever being able to reach total agreement on what the Bible says and means. In the face of this reality, we have to be patient with each other, and we have to learn to be humble, and never assume we have it all figured out.

The best scientists have learned to keep the same kind of humility. Despite faith they place in human reason, they realize that they cannot afford to assume that human reason is infallible or that human knowledge is complete. Science is an ongoing process in which old theories are constantly examined, modified, and replaced. Though undergraduate science textbooks might give the impression that all of the questions have been settled and scientists all agree with each other, a quick look through cutting-edge academic journals shows that this is not the case. Science, by its nature, is a continual work in progress.

Some critics of secular science say that scientific revolutions are really more about social forces than they are about facts. Thomas S. Kuhn's *The Structure of Scientific Revolutions*[26] is sometimes cited in discussions of this nature. There may be some truth to that statement, but it must not be overstated. The earth does not orbit the sun because of social forces. The interpretation some people assign to this state of affairs (i.e. Humans are not at the center of the universe, so they must not be very important.) may be values-laden, but the physical reality remains the same. Regardless of one's redeemed or fallen nature, political view, or feelings, the earth orbits the sun and not the other way around.

Challenge #5: Scientists believe reality is realistic, but not all Christians agree.

Scientists, of necessity, pretty much take the world at face value. I make this statement somewhat cautiously because quantum physics has ventured into some pretty strange realms of late. It is hard for a lay reader of some of the new books on subjects like string theory to tell where hard science ends and speculation begins. My assertion here is more basic: A scientist who finds a bone assumes it came from a real animal that lived sometime in the past. A scientist who sees stars through a telescope assumes that the stars were really there at the time they produced the light. If other galaxies are so far away that it takes millions of years for their light to reach us, the universe must be millions of years old, or we could not see them.

Some Christians have proposed that this assumption is false. Philip Henry Gosse, in the year 1857, responded to the early evidence for an ancient earth with one of the first examples of what I call the Illusion of Age theory. In his book, *Omphalos: An Attempt to Untie the Geological Knot,* he proposed that the

---

[26] Thomas S. Kuhn. *The Structure of Scientific Revolutions* (Chicago: The University of Chicago Press, 1962).

evidence of science did indeed point to an ancient planet, but that God had supernaturally created the earth with the appearance of age. Gosse viewed nature as a cycle that constantly repeats, like a continuously turning wheel. Because creation would be an interruption of the cycle, one might just as easily start with the chicken as the egg, metaphorically speaking. God could have created fully-grown trees just as easily as seeds and could have given Adam and Eve belly buttons (*Omphalos* is Greek for navel), even though they never had umbilical cords attached to them. The *real age* of the universe, in his theory, might only be a few thousand years whereas the *ideal age* would be ancient. An ancient world and a world created with the appearance of age would be indistinguishable. [27]

If you allow for the existence of an all-powerful God, as one of my friends points out, you allow for the possibility that God created all of us five minutes ago and gave us false memories of lives we never lived. The question is not whether an all-powerful God *could* create a world with a false history, but whether the God described in the Bible *does* operate that way. Some would say it places God in the role of a trickster who deliberately misleads people. Though there are stories in the Bible in which God allowed false prophets to produce counterfeit miracles in order to test the faith of the people, there is no biblical example of God or the devil creating false artifacts.

Besides all of that, some of the Bible's authors point to nature as evidence for God's reality. The Apostle Paul writes, in Romans 1:20, that God's invisible attributes are so evident in nature that no one has any excuse for not believing in him. "The heavens," according to Psalm 19:1, "declare the glory of God."

One benefit of Gosse's "illusion" theory, however, is that it neatly compartmentalizes science and religion. Science can

---

[27] Bernard Ramm, *The Christian View of Science and Scripture* (Grand Rapids: William B. Eerdmans Publishing Company, 1954), 133-134.

discover what it will while religion remains untouched. Some strict Bible literalists find this approach to be the cleanest way to resolve the apparent conflicts between the Bible and science. While others propose that secular scientists are either corrupt or incompetent, proponents of the illusion of age approach do not blame scientists for mistaking the ideal age of the universe for its real age.

Challenge #6: Scientific language is precise, but biblical language is nuanced.

Academic writers in both "hard" disciplines like physics and chemistry and "soft" disciplines like psychology and philosophy go to great pains to define their terms in precise, technical ways. The biblical writers, on the other hand, wrote using a variety of literary forms, and some wrote using cultural conventions that are unfamiliar to modern readers. This sometimes makes it difficult to draw exact parallels between the events described in biblical texts and the artifacts uncovered by archaeologists, especially when studying the events described in the Genesis creation narratives.

Many of the Greek and Hebrew words in the original texts also had multiple layers of meaning that were lost in translation. The names of characters, in particular, were often rich with symbolic meaning. I have tried, as much as possible, to correct for these linguistic and cultural gaps in my chapters on Genesis by referring to some of the leading scholarly commentaries. Many questions remain, of course, but the amount of data available to archaeologists has grown tremendously in recent years. In our next chapter, however, we will deal with the hardest challenge of the seven.

CHAPTER 2:
Science and the Bible: Tying a Rocket to a Cathedral

Challenge #7: Science is dispassionate about results (ideally, at least), but religion demands dedication to doctrine.

What happens when you tie a rocket to a cathedral? Unless you use a really long and flexible bungee cord, you will either crash the rocket or pull down the cathedral. Neither one can function the way it was meant to if they are bound too tightly together. Rockets are designed to move, and cathedrals are designed to be stationary.

This is one of the most serious challenges to Christian scientists and educators. The Bible doesn't change, but science changes constantly, and the discoveries of the past two centuries have been nothing short of staggering. Christian approaches to these issues are generally rooted in one of two basic assumptions:

Some Christians have decided, at the outset, that only a literal reading of Genesis is acceptable and that any evidence that does not support a literal view has either been misinterpreted or is some kind of illusion. Creationists who strongly argue for a 6,000-to-10,000-year-old universe and a planetary flood that created the planet's geological features fall into this group. Much of their effort is spent trying to discredit mainstream scientific theories. They use the language of science to argue for the inadequacy of science, and many of their apologists view mainstream scientists as enemies. There are notable exceptions of course, but the moderate voices of reason seldom get as much media attention as the more colorful representatives of this position. Kurt Wise and Todd Wood, mentioned in the previous chapter, are somewhat more positive in their treatment of science, but remain strict literalists in their reading of Genesis.

Other Christians proceed from the more general assumptions that all truth is God's truth, and that everything will make sense in the end. They are not greatly disturbed by apparent contradictions between a literal reading of Genesis and the discoveries of modern scientists. They recognize that science—by its nature—is a work in progress, and so is our understanding of the Bible. Old earth creationists and theistic evolutionists would fall into this group. B.B. Warfield, Bruce Waltke, and Augustine represent this point of view. (We'll meet them later in this chapter.)

Both groups believe the apparent contradictions between science and the Bible will ultimately work themselves out in the end, but they differ on how they see this happening. The first group believes the facts will ultimately support a literal reading of Genesis, and the second allows for a wider range of solutions. Reading the Bible, they would say, is not like reading the newspaper. They believe the Bible will ultimately be proven true in all the things it intends to teach, but sometimes we confuse the literary packaging with the message itself.

An axiom, according to *The Westminster Dictionary of Theological Terms*, is "a statement that requires no proof and thus serves as a premise or basis for arguments. In Christian belief, 'God exists' would be an axiom."[28] The belief (held by young earth creationists) that only a literal Bible is acceptable and the belief (held by both) that everything will ultimately work out in the end are axioms. They are unprovable faith statements that shape the ways Christians tackle the issues we'll be discussing here. An axiom, to put it in layman terms,

---

[28] Donald K. McKim *Westminster Dictionary of Theological Terms* (Louisville, KY: Westminster John Knox Press, 1996), 23, 24.

is *something that a person assumes just has to be true no matter what.*

Secular scientists have axioms of their own. The belief that everything we encounter must be the result of natural forces is a bedrock preconception many atheists hold, and any explanation that involves a personal God is eliminated at the outset. As noted earlier, there are many scientists who believe in God but limit themselves to the study of natural explanations in their work and leave the supernatural to theologians. Their personal belief systems differ from the ground rules they accept in their work environments.

Foundational beliefs like the ones mentioned above are harder to change than other elements of a person's belief system. They are deeply rooted and can neither be proved nor disproved by any particular argument or piece of evidence. They are "reasoned from" rather than "reasoned to." They are not completely impervious to evidence, however. Sometimes atheists become Christians, and sometimes professing Christians lose their faith.

By prescribing ahead of time what the evidence will be allowed to say, some strict literalist organizations have placed scientists in a delicate position. These organizations can be very hard on scientists whose work does not support their position, *but what if you are the scientist?* What if, as a Christian, you want the results of your work to support a literal reading of the Bible, but the evidence won't cooperate? Should you be honest about your discoveries or only report discoveries that support your worldview? Some scientists have found themselves in that position.

Chemist P. Edgar Hare was one of the founding members of the Seventh Day Adventists' Geoscience Research Institute. The organization was founded to promote young earth creationism. While attempting to show that marine animals found at different geological strata were actually the same age, Hare found convincing evidence that went against the organization's young earth assumptions and ended up resigning.[29]

The Creationist Research Society required its members to sign a manifesto affirming a literal seven-day creation week, a global flood, and biological change only within created kinds of animals. Larry Butler of Purdue University served as the organization's president at one point but later resigned when his biochemical research into the enzymes of various animals convinced him that there was a hereditary relationship between organisms as diverse as bacteria and snakes.[30]

Neither of these scientists, to my knowledge, lost his faith as a result of his discoveries. Stepping back from strict literalism is not necessarily a slippery slope leading to agnosticism. It has been for some people, however, and that is why some strict literalists guard their territory so fiercely. Their concern is easy enough to understand.

Some might say that these scientists, in their pride, placed human reason and experience above the Bible, but I disagree. They simply reported what they had found. In spite of their good faith efforts to provide evidence for a straightforward literal reading of the Bible, they were unable to find the kind of evidence they were looking for, and they admitted it. Candor about experiences like these is not always appreciated by Christian organizations. There are other examples.

Bill Dembski, a professor at Southwestern Theological Seminary in Fort Worth, Texas, was a leading spokesman for the Intelligent Design movement and the founder of the Discovery Institute. During the controversial *Kitzmiller v. the Dover School District* trial of 2004, Dembski fought to bring God back into science, so to speak, only to have his theories ridiculed by those holding opposing views. After enduring the attacks of evolutionists from the left, he has more recently had to deal with opposition from fellow evangelicals for statements he made in his book *The End of Christianity*, published in

---

[29] Denis Alexander, *Rebuilding the Matrix* (Grand Rapids: Zondervan, 2001), 307-310.

[30] Ibid., 309.

2009. Referring to the evidence for an ancient earth and the difficulties of defending a global flood, Dembski made the following statement:

> *The young earth solution to reconciling the order of creation with natural history makes good exegetical and theological sense. Indeed, the overwhelming consensus of theologians up through the Reformation held to this view. I myself would adopt it in a heartbeat except that nature seems to present such strong evidence against it.*[31]

Dembski was called before the seminary's president and board of directors, and he ended up, more or less, recanting statements he made in the book.[32] Interestingly enough, Page Patterson, the seminary's president, served as the managing editor of the Criswell Study Bible which allowed for the possibility of an old earth.[33] Some of Dembski's other conclusions, however, apparently placed him too far outside Patterson's comfort zone and that of the board.

Speaking candidly about the dilemma Christians face when it comes to dealing with the theory of evolution, Bruce Waltke, a highly respected Old Testament scholar posted the following comment on a website in 2010:

> *"...if the data is [sic] overwhelmingly in favor of evolution, to deny that reality will make us a cult...some odd group that is not really interacting*

---

[31] William Dembski, *The End of Christianity* (Nashville: B&H Academic, 2009), 55.

[32] David Roach, "How Old? Age of Earth Debated Among SBC Scholars," *Florida Baptist Witness*, Oct 20, 2010.

[33] W.A. Criswell, *The Criswell Study Bible* (Nashville: Zondervan, 1979), 3.

> with the world. And rightly so, because we are not using our gifts and trusting God's Providence that brought us to this point of our awareness."[34]

Waltke lost his job as a result of these comments, and has since been hired by another seminary.[35]

These situations bring up some uncomfortable questions: How far should Christians go in adapting their understanding of the Bible to the discoveries of modern science? At one extreme, we risk giving up too much and simply becoming mirrors of the societies we live in. At the other, we risk becoming like a flat earth society.

An important question to ask would be whether certain beliefs lie at the very core of the Christian faith, or whether they only represent contemporary understandings of it. Let's consider two examples.

Author John Dominic Crossan is one of the co-founders of the controversial Jesus Seminar. Crossan believes the disciples of Christ probably concocted the story of the bodily resurrection of Christ to establish their power base. It is more likely, to Crossan's way of thinking, that the body of Jesus was buried in a shallow grave, dug up, and eaten by dogs. The disciples may have had visions of Jesus, but no more than that. Interestingly enough, Crossan has a high regard for Jesus and considers himself a Christian. Apparently he defines a Christian as a person whose life has been touched by the story and teachings of Jesus in a way that causes him or her to live a better and more unselfish life. Belief in the miraculous aspects

---

[34] "OT Scholar Bruce Waltke Resigns Following Evolution Comments." *Christianity Today.* Posted on April 9, 2010. http://blog.christianitytoday.com/ctliveblog/archives/2010/04/ot_scholar_bruc.html

[35] Charles Honey, "Adamant on Adam," *Christianity Today*, posted: 5/25/10, http://www.christianitytoday.com/2010/june/1.14.html [accessed August 18, 2011].

of the story, for Crossan, is not required.[36] *Just who is to say, Crossan might ask, what someone has to believe to be called a Christian?* That is a good question, indeed.

We could answer by asking how the New Testament defines Christianity, how Roman Catholic, Eastern Orthodox, or Protestant churches have defined it through the ages, or what modern people mean when they use the word *Christian*. In the United Methodist Church my family attended, the basic beliefs of Christianity were summarized in our weekly recitation of the Apostle's Creed. The creed has roots in the early history of the church at a time when offshoots of Christianity were popping up because neither the canon of Scripture nor the doctrines of the church had been formalized.[37] In our own time, some evangelicals try to avoid the confusion over the meaning of the word Christian by calling themselves Christ-followers. Should we, in light of Crossan's comments, refer to ourselves as Followers of the Physically Resurrected Christ?" That would clear up the confusion, but it is a bit of a mouthful.

Takes on the Bible that deny the resurrection of Christ from the dead and treat heaven strictly as a metaphor, many of us would say, cut the living heart out of Christianity. The Apostle Paul wrote, "If we have hoped in Christ in this life only, we are of all men most to be pitied." (1 Corinthians 15:19, NASB) What good is a metaphorical heaven when death is a concrete reality? Though Christians may not agree about whether heaven's streets are paved with literal gold, most Christians consider belief in an afterlife to be a core tenet of Christianity. I definitely side with the majority on that issue. The point here, however, is not to discuss the importance of the doctrine of the

---

[36] Robert B. Stewart, *The Resurrection of Jesus* (Minneapolis, MN: Fortress Press, 2006), 11, 177-178, 186.

[37] Justo L. Gonzales, *The Story of Christianity*, Volume 1 (New York: HarperCollins, 2010), 79.

resurrection to Christianity, but to consider another question: how does one determine the core beliefs of Christianity?

If the issues discussed by Crossan and the Jesus Seminar strike at the very heart of the definition of what it means to be a Christian, there are other issues that were considered devisive in the past but are considered trivial today. One fascinating case study in the relationship between Christianity and science is the sixth-century "war of words" between Cosmas Indicopleustes, a Nestorian monk, and John Philoponus, a Christian philosopher from Alexandria. A debate had arisen in the church about the shape of the earth: Jews traditionally believed in a flat earth, whereas Greeks believed the earth was spherical. The Greeks were pagans but, as history has demonstrated, they were also better astronomers.

Philoponus wrote that belief in a round earth was perfectly compatible with the Genesis creation story, and Cosmas wrote a bitter rebuttal rejecting Philoponus' work as "two-faced" and hostile to a Christian view of the universe. Yes, the pagans had studied solar and lunar eclipses and the circular motion of the heavens, but true and faithful Christians would not compromise their faith by accepting such evidence. Philoponus responded that Cosmas' arguments were "the brayings of an ignorant ass" and that he hoped people outside the faith community would not consider them representative of Christianity as a whole. Cosmas' views were actually the minority position, and the church ultimately accepted the "round earth" cosmology of the Greeks. This, unfortunately, included some flawed ideas that cropped up later when the discoveries of Galileo and Copernicus showed that earth was not at the center of the universe.[38]

If belief in the Resurrection lies at one end of the theological continuum and the shape of the planet lies at the other, where

---

[38]Ken Keathley, "Flat or Round? The Sixth Century Debate Over the Shape of the Earth," in Stewart, ed. *Intelligent Design: William A. Dembski and Michael Ruse in Dialogue* (Minneapolis: Fortress Press, 2007), 198, 199.

GENESIS and the Thoughtful Christian/45

on that continuum do modern debates about issues like the age of the earth and human evolution lie? That is difficult to say. It is easy for us to laugh at flat earth Christians now, but we have the luxury of looking back from over a thousand years in the future. We have no easy way to predict how the Christians of future generations might view the issues we face today.

Some Christians in the U.S. might be surprised to discover that the "culture war" between evangelical Christianity and secularism is over a century old. Recent skirmishes are only the latest manifestations of a conflict that has been going on in Protestant, Catholic, and Jewish circles since the middle of the 1800s. Each of these groups experienced a kind of split between more progressive and more conservative branches.[39] For Protestant Christians, this led to a de-emphasis on the supernatural aspects of the Bible and an emphasis on doing good for society—what is commonly referred to as the "social gospel." Conservative Protestants responded in a number of ways. One was the publication of *The Fundamentals*, a twelve-volume set of writings by leading conservative scholars defining and defending what they believed were the core beliefs of Christianity. The popular terms *fundamentalist* and *fundamentalism* came from this set of books.[40]

Many modern readers might picture the authors of these books as fiery country preachers railing against the dangers of progress, but that picture fails to do justice to the reality. One of the authors was Princeton theologian B.B. Warfield, a leading advocate for the use of reason in defending the Christian faith[41] and a spokesman for the verbal theory of the

---

[39] James Davison Hunter, *Culture Wars* (New York: Basic Books, 1991), 77.

[40] Hunter, *Culture Wars*, 83; Hunter, Justo Gonzalez, *The Story of Christianity, Volume 2* (New York: HarperCollins, 2010), 342-343. C.T. McIntyre, "The Fundamentals," *Evangelical Dictionary of Theology*, ed. Walter Elwell (Grand Rapids: Baker Academic, 2001), 475-476.

[41] M.A. Noll. "Warfield, Benjamin Breckenridge," *Evangelical*

inspiration of Scripture.[42] B.B. Warfield's 1895 statement in *The Presbyterian Message* might surprise some readers:

> *The really pressing question with regard to the doctrine of evolution is not whether the old faith can live with this new doctrine.... We may be sure that the old faith will be able not merely to live with, but to assimilate to itself all facts.... The only living question with regard to the doctrine of evolution still is whether or not it is true.*[43]

These statements, from one of the men who wrote the book (one of them, at least) on fundamentalism, sound remarkably like the ones made by Bruce Waltke in 2010. Warfield's statement still leaves readers with unanswered questions. Just how far can the Bible's teachings be stretched to assimilate new discoveries before the Bible ceases to be the Bible?

Some readers might, understandably, take issue with Warfield's use of the word *doctrine* in reference to evolution, because it is the elevation of evolution almost to the status of a religious doctrine that troubles many of us. We have seen some evolutionists place chrome Darwin fishes on their cars in deliberate mockery of the chrome Jesus fishes Christians display. Secularists don't really worship Darwin, of course, but they do seem to consider his name to be an iconic expression of their worldview. Whatever it symbolizes to modern Americans,

---

Dictionary of Theology, 2nd Ed. ed. Walter Elwell (Grand Rapids: Baker Academic), 1257-1258 and John Newport, *Life's Ultimate Questions* (Dallas: Word Publishing, 1989), 140, 154, 417.

[42] I.S. Rennie. "Verbal Inspiration," *Evangelical Dictionary of Theology*, 2nd Ed. Walter Elwell, Ed. (Grand Rapids: Baker Academic), 1242-1244.

[43] Benjamin Warfield, "The Present Status of the Doctrine of Evolution." *The Presbyterian Message*, December 5, 1895, 7-8.

Warfield clearly did not view evolution as a religious doctrine. A brief look at a dictionary shows that the word has a range of meanings, including some so basic as, "something taught; teachings."[44]

As mentioned previously, the risks of intellectual isolationism, of becoming a "flat earth society," are weighed against the risks of assimilating too much and giving up core beliefs. If Christians are right, this risk could have eternal consequences. For Christian denominations, erring in either direction could have immediate and measurable consequences as well. In the mid-1800s, many of the mainline Protestant denominations chose to de-emphasize the supernatural aspects of Christianity and focus on reforming society. Since that time, many of these denominations have shrunk markedly while evangelical and charismatic groups have grown. According to one recent statistic, there are now more Muslims in the United States than there are Methodists, Episcopalians, and Presbyterians.[45] Though membership alone is not an indicator of truthfulness, the decline in the mainline denominations does bring up an important point: do people want a faith that mirrors the sensibilities of society or one that stands apart from society and challenges it?

N. J. Demerath, a University of Massachusetts sociologist, wrote about this phenomenon in "Cultural Victory and Organizational Defeat in the Paradoxical Decline of Liberal Protestantism." The core values of liberal Protestantism include pluralism, individualism, tolerance, free critical inquiry, and the authority of human experience. Strong organizations, he argues, are not built on diversity, autonomy, individualism, and criticism. The success of these organizations, he believes, planted the seeds for their decline.[46]

---

[44] *Webster's New World Dictionary of the American Language.* Second College Edition (New York: Prentice Hall, 1986), 414.

[45] Hunter, 73, and Gonzalez, 492-493.

[46] N. Jay Demerath, "Cultural Victory and Organizational Defeat in the

Demerath's statement leaves us with much to think about. In a world that has been shrunken by advances in travel and communication technology, many of the values he listed are necessities for survival. Some writers differentiate, however, between the idea of pluralism as a fact of life and pluralism as an ideology. People might have an equal right to hold their beliefs, but does that mean *the beliefs themselves* have equal warrant? Dedicated pluralists say yes, but are they contradicting themselves by doing so? Pluralism is, after all, a belief system. Is it just as valid to be an exclusivist as it is to be a pluralist? A certain amount of free discussion in an organization, a "medicinal dose of anarchy," keeps a religious group healthy, but too much freedom to believe whatever one wants to ultimately brings it down. Those who stand for everything stand for nothing, as the saying goes.

What about the emphasis on reforming society at the expense of Christianity's supernatural elements? Isn't reforming society a worthy goal? Author Philip Yancey once wrote about the "risks of relevance," citing the example of the poet T.S. Eliot, who gave up writing poetry for a while after becoming a Christian and wrote books on economics and sociology instead. Poetry, to the newly converted Eliot, seemed frivolous and unimportant compared to the political and social issues of his day. If he was going to make a difference, he thought, he needed to write about what mattered. The political issues of Eliot's day have largely been forgotten, and his books on economics are out of print, but his poetry is still assigned reading in college literature classes, and major bookstores still sell collections of his poems.[47] How, then, should Christians and Christian institutions deal with the "risk of relevance?"

Paradoxical Decline of Liberal Protestantism," *Journal for the Scientific Study of Religion*, 34(4) (1995), 458-469, and John Murray Cuddihy, *No Offense: Civil Religion and Protestant Taste* (New York: Seabury, 1978).

[47] Philip Yancey, *I Was Just Wondering* (Grand Rapids: William B. Eerdmans Publishing, 1989), 129-135.

Remaining frozen in time does not seem like the answer, but we can go so far in adapting ourselves to the needs of one generation that we have nothing to say to the next.

My own Southern Baptist denomination has struggled between fundamentalism and progressivism for more than a century. The latest upheaval, referred to by some as the Conservative Resurgence and others as the Fundamentalist Takeover, led to a bitter split with fundamentalists on the far right, liberals on the far left, and everyone else caught somewhere in the middle. Financial support of the Cooperative Program, the mechanism that funds the large denomination's educational institutions, missions programs, and charitable work, dropped by about 33 percent between 1984 and 2004.[48] (Correlation does not necessarily indicate causality, however. Though the conflict is probably part of the picture, changes in the spending habits of American Christians is bound to have an influence on their charitable giving as well.)

After reading about the Myers-Briggs personality types, I have often wondered what role basic personality differences played in the internal culture war that divided my denomination. Some people, by nature, ask questions and explore possibilities, and others doggedly defend tradition. The traditionalists give a group its stability, and the explorers give it the ability to creatively adapt to new situations. Both are essential to the survival of a healthy organization but, all too often, they end up demonizing each other. The explorers are "flighty liberals," and the traditionalists are "stuck in the dark ages."

I am concerned that what evangelicals are doing now is not working. According to a study by the Barna Research Group, approximately 43% of the young Christians who are active in church as teenagers drop out as they enter adulthood. Some common complaints were that they found their churches to be fearful of people with other beliefs, unwilling to discuss

---

[48] Chad Owen Brand and David E. Hankins, *One Sacred Effort* (Nashville: Broadman and Holman, 2005), 161.

difficult issues, anti-pleasure, too quick to demonize those outside of the church, and/or—particularly relevant to our subject—they were anti-science.[49]

If too much emphasis on diversity and free critical inquiry helped to bring about the decline in membership of once-thriving denominations like the Methodists (my parents' denomination), but too much rigidity and isolationism is toxic to the faith of the young Christians described in the Barna survey, what is the solution? How do we preserve denominational stability without squelching or ignoring honest questions about the relationship of orthodox Christianity (however defined by the denomination in question) to the issues of the day? How, for purposes of the subject at hand, do we allow discoveries in biology, geology, archaeology, astronomy, and other fields to inform our understanding of the Bible without losing our uniquely Christian outlook on the world? Answering these questions on an individual level is difficult enough, but responding to them as denominational groups is even more difficult. One phrase Southern Baptist leaders employ is "unity in essentials, diversity in nonessentials."[50] If we could ever reach consensus on just what the essentials are, perhaps we could learn to get along. Apostles' Creed, anyone?

In my talk of preserving Christian orthodoxy, I've assumed we have something worth preserving, and I realize some of my readers might not share that assumption. "Perhaps," some might say, "we've been wrong for centuries, and it's time to let it all go." In response, I have tried asking myself questions like, *Why do I believe this, anyway?* or *If I didn't already believe this, what would it take to convince me?* but only an outsider could say whether my answers are convincing. I'm only guessing. All I can do is lay my stories and research on the

---

[49] David Kinnaman, *You Lost Me* (Grand Rapids: BakerBooks, 2011).

[50] Brand and Hankins, *One Sacred Effort*, 81.

table, then leave it up to you, the reader, to decide what you think.

In our discussion about science, history, and denominational politics, however, we must not lose sight of one basic assumption: Christians assume their faith will survive because God is real. The late apologist Francis Schaeffer entitled one of his books *He is There and He is Not Silent*, and that is the conviction of many Christians. No amount of logical argument can prop up a faith that is ultimately rooted in falsehood, but if God is really real, we can only hope he has not given up on Western civilization and moved on to more fertile ground. The staggering growth of Christianity in places like Africa and South Korea is encouraging to Christians in the West, but we are also saddened by empty cathedrals in countries like England where Christians are now a minority.

Interestingly, the questions surrounding the book of Genesis did not all begin with the advent of modern science. St. Augustine wrote *Literal Commentary on Genesis* in the late fourth and early fifth centuries. He included the following comment for his readers:

> *...different interpretations are sometimes possible without prejudice to the faith we have received. In such a case, we should not rush in headlong and so firmly take our stand on one side that, if further progress in the search for truth justly undermines this position, we, too, fall with it.*[51]

In our next few chapters, we'll take guided tours of two fascinating timelines. You'll read about the origin of the universe, the origin of life, the origin of humanity, and the origins of human civilization. You will also read about the disasters that nearly swept humanity from the earth. The

---

[51] Randy Moore, Mark Decker, and Sehoya Cotner, *Chronology of the Evolution-Creation Controversy* (Santa Barbara, CA: ABC-CLIO, LLC, 2010), 7.

biblical and scientific timelines, read back to back, are surprisingly similar in some ways, yet startlingly different in others.

Some would consider the biblical story to be quaint, outdated, and maybe even offensive in places. The scientific story, for them, is the only story worthy of sophisticated, modern people. Others would consider the biblical account to be the "true science" and the scientific account to be a sad counterfeit put forth by a skeptical culture.

Still others would find both stories compelling. Comparing them, they might say, is like comparing a musical composition to a painting or an orchestra's rendition of a storm to a weather report. The differences in the stories, for them, come largely from the differences inherent in the media used to express them.

As we read the chapters that follow, we might imagine ourselves walking through two museums. In the first, we see paintings, ancient documents, reproductions of archaeological dig sites, and pottery. In the second, we see images from the Hubble telescope, the restored skeletons of prehistoric beasts and hominids, paintings copied from the walls of smoky caves, and models of DNA helixes. The past awaits us, so let's start at the beginning.

CHAPTER 3:
A Guided Tour of Genesis, Part I: Losing Paradise

The biblical story of the development of the world and humanity is fairly short and it appears, at first glance anyway, to be fairly simple. It begins with the description of a week in which God created the heavens and the earth. The universe begins as a primeval, watery chaos.

Several points could be made here. Notice, first of all, how our modern creation narratives start in outer space, but the Hebrew narrative starts with an empty, chaotic ocean. As one author expressed it, modern science begins with a Big Bang, whereas Genesis begins with a Big Splash.[52] The Big Bang is the sort of thing you'd expect from a group of people conscious of the fact that we live on a round planet orbiting a star in a galaxy. The ancient Hebrews would not have thought of the creation in that way. The beginning of Genesis reminds me more of the way James Michner started the novel *Hawaii,* with the Hawaiian Islands being belched up out of the ocean by a volcanic eruption.[53] Interestingly enough, a description of the world's beginning in Proverbs 8:27 pictures God inscribing a circle on the face of the deep. Even though both the writer and the original audience probably envisioned a disk-shaped world floating on a bottomless ocean, the description manages to

---

[52] Conrad Hyers, *The Meaning of Creation* (Atlanta: John Knox Press, 1984), 39.

[53] James A. Michener, *Hawaii* (New York: Bantam, 1957), 1-15.

sound surprisingly modern. Modern writers still refer to the earth as a tiny island in the ocean of space.

God's spirit enters the primeval chaos and moves "on the face of the waters." He starts by putting things in order. In days one through three, respectively, God separates day from night, water from air, and land from sea. During day three, he also creates green plants. In the first three days, you'll notice, God is setting up empty stages. At the end of those days, the day and night, the water and air, and the land and trees are still unoccupied. In days four through six, respectively, God fills the day and night with celestial bodies, fills the air with birds and the sea with the "great sea monsters," and populates the earth with "creeping things" (presumably reptiles and amphibians), land animals, and, last of all, male and female humans (Genesis 1:1-2:4a). At the end of this sequence, God declares his creation good and commands his new creations to be fruitful and multiply. (Genesis 1:28) Though some have tried to link Eden's forbidden fruit with sexuality, there is no hint here that God finds "multiplication" to be sinful.

The creation week ends with a seventh day in which God rests from his work of creation. God sets this day aside as a holy day, and it becomes the basis for the Jewish sabbath. Gary Rendsburg, a Jewish scholar, compares God's sanctification of the sabbath to the end of the *Enuma Elish*, a Babylonian creation epic, in which the sun god Marduk sanctifies his temple in Babylon. In Genesis, God makes the entire world his temple and sets aside a period of time rather than a physical place.[54] Some scholars believe this part of Genesis may have been written at the time the Hebrew people were in exile in Babylon. Compared to their great cities and the mighty temple there, the Jews might have felt that what they had was inferior. Genesis provides a response to that: "They've got a temple, but our God inhabits the whole of creation. That's his temple."

---

[54] Gary A. Rendsburg, "Lecture 2: Genesis 1," *The Book of Genesis.* The Great Courses audio lecture series, produced by The Teaching Company of Chantilly, VA. 2006.

Many Christians throughout the history of the church have taken every aspect of this narrative literally, and some believe it advocates a recently created universe. In the seventeenth century, Archbishop James Ussher of Scotland dated the world's creation to the year 4004 B.C., and some groups still adhere to Ussher's dating.[55] Others, as early as Augustine, have interpreted the narrative somewhat more fluidly,[56] citing 2 Peter 2:8: "...with the Lord, one day is as a thousand years and a thousand years as one day." In the Book of Enoch, a book of Jewish apocalyptic fiction written during the period between the Christian Old and New Testaments, God describes his creative acts to Enoch as follows:

And I assigned the eighth day to be the first day created after my work, and the first seven to revolve in the form of seven thousand. At the beginning of the eighth thousand I appointed an uncounted time, an infinity, unmeasured by years, months, weeks, days or hours.[57]

This piece of literature is, as I have said, not part of the Bible. What it does show, however, is that the idea of the seven days of creation being longer than twenty-four hours is not entirely confined to the modern age.

Both of the previous theories treat the creation week event as a literal description. Those who read the days as ages often note the fluidity of the Hebrew word *yom* that can refer to either a day or a longer period of time much the way that we

---

[55] John P. Newport, *Life's Ultimate Questions* (Dallas: Word Publishing, 1989), 148.

[4] Hugh Ross, *Creation and Time* (Colorado Spirngs: Navpress, 1994), 19-20.

[57] The Book of the Secrets of Enoch quoted in Willis Barnstone, *The Other Bible* (New York: HarperCollins, 2005), 7.

would refer to "the day of the steam engine." Reading the book as literal history is not the only option, however. Many readers have noted the three days of separating and the three days of filling and view the creation week narrative primarily as theological poetry. Any correspondence to the real order of creation, for them, is secondary to the logic of the poetry and trying to scientifically or miraculously explain why the sun doesn't show up until the fourth day misses the point. The important part, for them, is not the structure but the theology. Certainly the habitat had to come into being before the animals could live there, they would say, but it wasn't meant to be read like a zoology textbook. People weren't writing many zoology textbooks in those days, but every civilization had poetic epics about its gods. For them, it was more important to be colorful than to be scientific. Compared to other creation stories of that time, however, Genesis is very restrained in its imagery. There are no gods cutting each other's heads off or bleeding the oceans into existence.

Some have suggested that the late placement of the sun, moon, and stars could have had a spiritual purpose rather than a historical one. Sun worship was widespread, and placing the creation of the sun at the beginning of the narrative could have served to reinforce sun worship. The narrative refuses even to name the sun and moon, referring to them as the "greater light" and the "lesser light" because the names of both bodies were the personal names of pagan deities.[58] Though these were not the names of the Mesopotamian deities, imagine the confusion that would have taken place if the author had written something like, "And God created Marduk to rule the day and Diana to rule the night."

After the creation week, the Genesis story continues with the Garden of Eden narrative. Though fused together into a single narrative now, many believe the Creation Week and Garden of Eden stories once existed separately.[59] At the very

---

[58] Conrad Hyers, *The Meaning of Creation: Genesis and Modern Science* (Atlanta, John Knox Press, 1984), 21.

least, the differences in the narratives represent a shift in focus from the creation as a whole to humanity in particular. In the Eden narrative, God creates Adam from clay and creates a habitat, a garden, for him to dwell in. The name "Adam" means man or mankind.[60] It comes from the same basic root as *Edom,* which means "red" and refers to red clay. The name Eden refers to the area in which the garden is located and not to the garden itself. The world outside the garden, presumably, is barren.

God creates the animals and brings them to Adam. He names them but finds no suitable mate for himself. So God causes a deep sleep to fall upon Adam. He takes a rib from Adam's side and fashions a woman who will be his helpmate. The man and his wife live happily together for a while, until a serpent deceives the woman into eating of the forbidden Tree of the Knowledge of Good and Evil. Adam, enticed by his wife, eats of the same fruit. The two are expelled from the garden as punishment (Genesis 1:4-3:24). Adam, whose name is derived from a word meaning ground, must try to grow plants from cursed ground. Eve, whose name is derived from a word meaning "to live," must bear children in pain.

Again, many Christians have taken this story as literal history. Adam's name is found in the genealogies of historical figures including Christ (Luke 3:23-38). Paul, in Romans 5:14,15, writes of the reign of sin from Adam to Moses and describes Adam as a "type of him who was to come," a kind of negative Christ-symbol. If Adam's sin could corrupt the world, Christ's righteousness could save it. The physical location of Eden is described in detail, but pinpointing an exact location is difficult because two of the rivers, the Pishon and Gihon, are not on modern maps. Geologist Carol Hill, in "Eden, a Modern

---

[70]William S. Lasor, David A. Hubbard, and Frederic W. Bush, *Old Testament Survey* (Grand Rapids: Eerdmans Publishing Co, 1996), 10, 11.

[60] Ibid., 19.

Landscape," uses the details of the story to place Eden just north of the Persian Gulf.[61] Other authors have placed it in the mountains near the Caspian Sea or even in the region of the Blue Nile.[62]

Though elements like the details of the landscape point to at least a semi- literal interpretation, other elements of the narrative point to a more symbolic interpretation. Though Christians identify the serpent with Satan, the narrative itself does not supply this linkage. The book of Revelation (Revelation 20:2), in the Christian New Testament, refers to Satan as an ancient serpent, but the narrative itself never says that the serpent is Satan or that Satan entered into the serpent. He is simply an intelligent, talking reptile. This suggests to many that the serpent and other elements of the story are to be taken as pictorial symbols rather than concrete realities. This does not necessarily mean, however, that Genesis is completely devoid of historical meaning.

When Revelation 12:1-6 describes the birth of Christ, it refers to a woman "clothed in the sun, and the moon under her feet, and on her head a crown of twelve stars." This woman is giving birth to a male child, and a dragon waits to eat it as soon as it is born. Using colorful symbols presumably familiar to his audience, the author is describing a real event with historical characters, but the imagery he uses is very different from the straight historical narrative one finds in Matthew 1:18-2:23 and Luke 2:1-20. It is apparently a literal, historical event (presumably Christ being born and almost being killed by soldiers, though is not the only possible interpretation of the imagery) described in a decidedly non-literal way. Thus, even a

---

[61] Carol A. Hill, "The Garden of Eden: A Modern Landscape," *Perspectives on Science and Christian Faith*, 52 (March 2000), 31-46.

[62] Thomas V. Brisco, *Holman Bible Atlas* (Nashville: Holman, 1998), 33.

conservative reading of the Eden narrative leaves room for some symbolism.

The discovery of an ancient library in Ninevah in the mid-1800s was a boon to biblical scholars who wanted to understand the literature and culture of the world from which Abraham emerged. It also led to a great deal of controversy. There is a scene in *The Epic of Gilgamesh* in which a serpent steals the fruit of rejuvenation from Gilgamesh, eats it, sheds its skin, and crawls away.[63] Because the oldest versions of this story come from a time before the Hebrew language even existed, this scene has created a chicken-and-egg controversy about the imagery in Genesis. Did the Genesis prophet (Tradition says it was Moses, but Genesis itself never identifies its author.) use popularly recognized symbols in fashioning his theological history, or did the Gilgamesh epic go back to an earlier narrative that was closer to the biblical one? Where and when did the the "snake story" really originate?

Other parallels between the Bible and the Gilgamesh story, especially those between the stories of Noah and Utnapishtim, who both built arks and filled them with animals to survive a great flood,[64] suggest a common cultural heritage between the biblical author and the author of Gilgamesh. Some critics of Christianity have suggested that the authors of the Bible, far from being inspired by God, simply plagiarized their stories from existing legends. Conrad Hyers, like many Bible scholars, believes the author (or authors) of Genesis constructed their narratives to answer the challenges the Israelites faced as believers in one God living in the midst of many pagan civilizations.[65] They weren't copying the myths of their

---

[63] Maureen Gallery Kovacs, *The Epic of Gilgamesh* (Stanford: Stanford University Press, 1989), 107.

[64] Werner Keller, *The Bible as History* (New York: Bantam, 1982), 31-37.

[65] Conrad Hyers, *The Meaning of Creation* (Atlanta: John Knox Press, 1984).

neighbors and hoping nobody would notice. They were going head-to-head with them by building a monotheistic (single God) response to them. The message of Genesis to the Hebrews, like the title of a popular worship chorus, was "There is no God like our God." Some of these issues will appear again in later chapters. Suffice it to say that whatever similarities might exist between the narratives, there are notable differences as well. The two narratives have common elements, but they communicate very different messages.

Returning to the story, it is unclear how long Adam and Eve dwelled peacefully in Eden. It could have been days, weeks, or years. God's only condition for living there was that they never eat fruit from the Tree of the Knowledge of Good and Evil. "On the day you eat of it," he sternly warned Adam, "you will die" (Genesis 2:17). If this scene had appeared in a movie, viewers would have known that somebody was going to violate that rule before the movie ended. It was just a matter of time. As with other elements of the story, we are left with questions about the nature of the fruit. Was there really a piece of fruit, or was the fruit simply a metaphorical representation of the knowledge of good and evil? Why was that knowledge viewed negatively? If it had been called the Fruit of Evil or the Fruit of Godless Sensuality, or if the narrative—like the Greek story of Pandora's Box—had described the unleashing of destructive forces into the world, it would be easier to understand exactly what readers were meant to draw from the narrative. As it is, theologians hold a variety of opinions on this issue. As all readers know, the serpent tricked Eve into disobeying God's command, and Eve talked Adam into violating the command along with her. Eve often receives the blame for what happened, but Adam was also a willing participant. The passage seems to indicate that he watched the entire exchange between Eve and the serpent and said nothing.

As an immediate consequence of their sin, Adam and Eve become aware of their nudity and begin to feel shame. Why would this have happened? Partial or complete nudity is fairly common among some tribal cultures, and people aren't particularly disturbed by it. Isn't nudity acceptable between husbands and wives? God did, after all, command them to be fruitful and multiply. Christian theologians, through the centuries, have gone back and forth on their views of sexuality. Some have treated all sexuality as something evil, and others view it as one of God's good gifts that has become corrupted. For the first group, the eating of the fruit made them aware of the inherent sinfulness of sexuality. The second group would say it corrupted God's good gift by removing the innocence from sexuality and introducing shame into something that was once beautiful.

God had told Adam that he would die on the day he violated God's command not to eat from the tree, but we know, having read the rest of the story, that Adam did not die that day but actually lived to be quite old. How can we explain the apparent contradiction between what God said he would do and what he actually did? Did "day" not mean day, did "die" not mean die, or are we missing something else?

One simple answer is that Adam and Eve died spiritually on the day of their sin, and that physical death was an eventual consequence. Others point out that God sentenced them to death when he exiled them from Eden. He had cut off their access to the tree of life, a tree with fruit that could have made them immortal.[66]

Adam, Eve, and the serpent all received punishments that were appropriate to them. The serpent was condemned to crawl upon the ground. It would nip at the heels of Eve's children, and they would crush its head with stones. Adam was condemned to toil in the dirt to grow his food, and Eve would suffer as she gave birth to children. She would also be subservient to her husband.

---

[66] For a detailed review of the issues, see: Victor Hamilton, *The Book of Genesis, Chapters 1-17*, vol. 1 of *The New International Commentary on the Old Testament* (Grand Rapids: William B. Eerdmans Publishing Company, 1990), 173.

Christian theologians through the centuries have constructed elaborate theologies about the effects of the Fall on the physical and spiritual condition of humanity. Many of these spring not from the Genesis account itself, but from the writings of the apostle Paul in the books of Romans and 1 Corinthians. In Romans 5:12, Paul writes that sin and death entered the world through Adam, just as life entered the world through Christ. Many theologians have taken this to mean that there was no physical death at all before Adam's sin and that Adam, Eve, and their children would have been physically immortal if they had not defied God. Some have argued that belief in the existence of dinosaurs and other extinct creatures that lived and died before humans ever came on the scene contradicts this verse and undermines the very foundations of the Christian faith.

Again, the easiest way to resolve the apparent conflict is to argue that the death referred to in Romans 5:12 is, first and foremost, spiritual in nature. We already noted that Adam and Eve did not die physically the day they violated God's command, but they no longer enjoyed the fellowship with God they had enjoyed in Eden. That bond was broken. There is probably no need to point this out, but sin is a uniquely human problem. Dinosaurs and bacteria would not have had spirits in the human sense and would not have been capable of willfully defying God. The Greek word *cosma*, translated "world," can also refer explicitly to the human world (as in "God so loved the world...."), and the passages following Paul's statement in Romans 5 tell how Jesus saved humanity, not the animal world, from the consequences of sin.

What about physical death, though? We might be tempted to assume Adam and Eve were physically immortal before they ate of the fruit, and that their children would have been too. We might even assume that eating the fruit altered them physiologically. Genesis itself never actually says any of these things. It certainly doesn't include clinical references to aging genes and telomeres[67]. As far as the story is concerned, it would appear that Adam and Eve died physically as a result of their expulsion from Eden. By exiling them from Eden, God cut Adam and Eve and their children off from both the tree of life and from his own healing presence. That is all that can be found in the story itself.

The penalties Adam, Eve, and the serpent received raise questions as well. Are we to infer that God created thorns, altered female reproductive biology (or gave babies bigger heads), and removed the legs from snakes (Boa constrictors still have hip bones.) the day he pronounced judgment on them or were these punishments also consequences of being exiled from Eden? Adam, we might reason, lost the rich soil of the garden and had to farm in the wilderness beyond. The serpent, similarly, lost its home in the trees of the garden and had to crawl through the hot sands of the desert. Unless the plants in Eden had medicinal properties, Eve's increased pain in childbirth is harder to understand from a physical standpoint. It makes poetic sense because, like Adam's sentence, it matches the meaning of her name and the prescribed social role of her gender in Hebrew society. Admittedly, it does not yield itself to easy explanation otherwise.

Assuming the details were entirely literal, would Adam and Eve have had pain receptors prior to the Fall (to prevent them from harming themselves), or would they have had no need of them? Would they have had perfect immune systems or would they have had no need for immune systems? What would have happened in Eden's perfect world if Adam had fallen off a cliff or scraped against a sharp rock? I didn't think of any of these questions until I tried to write a novel about Adam and Eve and encountered all of the practical questions of living in a world without pain. Placing the characters in some kind of holographic simulation would have made it easier to write than putting them in a world with living biology.

---

[67] Telomeres are fibers found in cells that grow shorter every time the cell reproduces. When they get below a certain length, the cell no longer reproduces.

Some have used Paul's statement in Romans 8:19-22 to link all natural evil (i.e. hurricanes, earthquakes, illness) to the Fall. In those verses, Paul writes that the Creation eagerly awaits the revealing of the sons of God and groans for renewal. Neither these verses nor those in Genesis specifically connect natural evils like earthquakes and forest fires to Adam's sin, but many theologians over the centuries have made that connection. It is often said that we live in a fallen world. From the standpoint of biblical theology and everyday experience, it is hard to argue otherwise. Acts of nature are somewhat harder to peg theologically. Should we say that nature has fallen from grace or simply that it is still awaiting redemption?

Some painful natural evils like cancer seem to be the result of glitches that have built up in nature's system, and we long for God to step in and "debug" our genetic code. Thorns on roses are painful to us, but they also help plants protect themselves, so should we consider thorns to be a glitch, or are they somehow part of God's providence? What about earthquakes? Seismic activity knocks down our buildings, but it's also the planet's way of renewing itself. According to Ward and Brownlee, the authors of *Rare Earth*, plate tectonics are responsible for regulating the build-up of carbon dioxide in the earth's atmosphere and keep the oceans from filling in with silt. The sliding plates are also part of the mechanism that produces the magnetic field that surrounds the earth and screens out deadly cosmic radiation from space. Not one of the other planets in the solar system has plate tectonics, and thus they're all dead worlds.[68] Much of our theology about the effects of Adam's sin on nature is highly speculative, and Genesis gives us little to work with.

Environmentalists would have no trouble agreeing wholeheartedly with Paul's statement in Romans, but they would say that it is not the natural world that is fallen but humanity. They would say the natural world groans under the

---

[68] Peter D. Ward and Donald Brownlee, *Rare Earth* (New York: Copernicus, 2000), 191-230.

burden of human greed, exploitation, and waste. Adam represents the beginning of human intelligence and human civilization. The plight of his children, both before and after the Flood, shows what happens when human intelligence and human civilization go awry. Perhaps they have a point. I don't know if environmental damage was on Paul's mind when he wrote Romans 8:19-22. Would a 1st-century writer like Paul have shared the concerns of modern environmentalists? It is hard to say, but with modern concerns like pollution and deforestation, the passage about nature groaning for renewal rings with prophetic meaning.

In relating natural evils to the effects of the Fall, some authors like Bruce Waltke differentiate between moral evil and conditions that are hostile to human life. Murder, for example, is a case of moral evil and hurricanes and earthquakes are examples of "natural evil." (Some have referred to these, interestingly enough, as "acts of God.") He refers to these as *surd* evil. Genesis, he says, deals with the issue of moral evil, but books like Job testify that surd evil is a mystery.[69]

Perhaps the simplest explanation, assuming a historical Adam and Eve, is not that God altered the fundamental makeup of the natural world but that he removed his hedge of protection from these beings he had fashioned in his image. No longer "spoiled rich kids" living on the estate of their doting Father, Adam and Eve were on their own, at the full mercy of the natural world—almost. Though the relationship would never be the same, God never stopped watching after them. Adam and Eve's brief stint in paradise would only serve as a prologue for a drama that would fill the rest of the pages of the Bible: the story of God's efforts to redeem humanity.

Christians believe God's ultimate plan to save humanity was to assume human form in Christ and offer himself as a ransom for humanity. Through the centuries, a number of Christian writers have seen a foreshadow of this redemptive plan in

---

[69] Bruce K. Waltke, *Genesis: A Commentary* (Grand Rapids: Zondervan, 2001), 68, 69.

God's statement in Genesis 3:15 that Eve's "seed" would crush the serpent's head and the serpent would bruise his heel. Christian commentators as far back as Justin (in A.D. 60) and Irenaeus (A.D. 180) have referred to the passage as the *protoevangeleon* or first gospel. Interestingly, the idea that Genesis 3:15 predicts Messiah's victory over Satan predates the birth of Jesus and the beginning of the Christian era. It is found in Jewish writings including the Septuagint (a Greek translation of the Old Testament), the Palestinian targums (Jewish commentaries on Scripture), and possibly the Onqelos targum dating back to the third century B.C. Gordon Wenham, the author of *Word Biblical Commentary* on Genesis 1-15, does not believe the original narrator intended to write the passage as a messianic prophecy, but later writers read it in that light. This is an example of what theologians refer to as a *sensus plenior*, a deeper meaning not intended by the original author but intended by God. It violates the rules of interpretation used by modern Bible scholars but was apparently a common practice among the rabbis of that era.[70]

---

[70] Gordon J. Wenham, Genesis 1-5, vol. 1 of the *Word Biblical Commentary* (Nashville: Thomas Nelson, 1987), 80, 81.

CHAPTER 4:
A Guided Tour of Genesis, Part II: The Birth of Civilization

The story continues as Adam and Eve, forever barred from Eden, settle in the wilderness. Using skills they learned in Eden, they continue to farm and begin to bear children. Their first son is named Cain. Many think the name, *Qayin,* comes from the Hebrew *qanui,* "I have gained," because, as the text explains, his mother said, "I have gotten a man from God." The name may have further significance as "getting" becomes the theme of Cain's life and that of the civilization he later starts. Others have connected his name to a word meaning "smith" or "metalworker." Tubal Cain, one of his descendants (Genesis 5:12-14), worked in metal.[71]

Abel's name mean's "vapor." Does this represent his fleeting life or his spiritual nature? Wenham believes the name might have referred to the fleeting nature of human existence.[72] Perhaps he was a sickly infant that nearly died during infancy. (I have not encountered that theory in any of the commentaries.) The significance of his name is unclear, but Abel's is the tragic story of a righteous man murdered in his prime. Cain is a planter, and Abel herds sheep. Both bring offerings to God, but only Abel's is accepted. The story never

---

[71] Gordon J. Wenham, *Genesis 1-5*, vol. 1 of the *Word Biblical Commentary* (Nashville: Thomas Nelson, 1987), 101.

[72] Ibid., 102.

explains how they knew which offerings God had accepted and which he had rejected, and it is not entirely clear why God rejected Cain's offering or how Cain knew it had been rejected. Some have suggested, comparing Cain's sacrifice to those found in other parts of the Bible, that only a blood sacrifice would have been acceptable for God.[73] Perhaps Cain pridefully believed his plants, the fruit of his labor, were just as acceptable as his brother's animal sacrifices. The narrative itself is not specific on that score, however, and theologians continue to debate the issue.

Regardless of the reason for God's rejection, Cain becomes jealous, murders his brother Abel, and is cursed to wander the earth. God places a mark on him so that those he encounters in his exile will not kill him. This episode brings up even more questions: what was the mark, and who were the other people Cain feared? Either Adam and Eve had other children not recorded in the narrative, or God had, at some point, created other people we are not told about. Some have suggested that the story of Cain and Abel started out separate from the Adam and Eve story and was later fused with it.[74] Others have suggested that Adam was not, in fact, the first anatomically modern human being but a special man from whose lineage the nation of Israel and Christ would eventually emerge, and that other humans, "Pre-Adamites," shared the planet with him.[75] Ralph Elliott, in his controversial volume on Genesis, suggested that God created many men and women on the sixth day of creation and that the story shifts its focus to a single

---

[73] See Wenham, *Word Biblical Commentary, Volume 1*, 104 for a discussion of the various theories.

[74] Victor Hamilton, *The Book of Genesis, Chapters 1-17*, vol. 1 of *The New International Commentary of the Old Testament* (Grand Rapids: William B. Eerdmans Publishing Company, 1990), 233.

[75] Bernard Ramm, *The Christian View of Science and Scripture* (Eerdman's Publishing, 1951), 137, 143, 251.

couple to explain the origin of moral evil[76] (Genesis 4:1-4:16). Though the origin of Cain's wife is never explained, Cain marries and becomes the father of a civilization. Cain's children build a city (Genesis 4:17) and develop animal husbandry (Genesis 4:20), musical instruments (Genesis 4:21), and metallurgy (Genesis 4:22). One cannot help but wonder whether it is significant that all of these developments are linked to Cain's fallen line. Did the ancient Hebrews link civilization with evil, especially when they encountered advanced civilizations like Egypt and Babylon that did not share their beliefs? We will return to this later.

As the story continues, Adam and Eve have another son, Seth, who replaces Abel. The need for Seth to replace Abel might be seen as an argument against the traditional view that Adam and Eve had many other children by that time, unless they were all female or unless Cain had led the other male heirs astray. The author contrasts Seth's righteous line to the corrupt line of Cain. The genealogies of Adam, Seth, and their children indicate life spans of around 900 years (Genesis 5). The ages of Cain's children are not given, but most assume their lives were similarly long.

In Carol Hill's article "Making Sense of the Numbers in Genesis," she surveys the various theories about the advanced ages of the characters. One theory, popular among some groups, is the vapor canopy theory. This theory holds that the primeval atmosphere was shrouded in heavy layers of vapor that screened out cosmic radiation and lengthened human and animal life spans. She also explores a variety of other theories, including the theory that the ages were given in lunar months rather than years. Her own explanation is rooted in the Mesopotamian use of numbers as religious symbols built around a sexagesimal (base-60) numbering system. Hill also noted the use of number symbolism in Egyptian culture where

---

[76] Ralph H. Elliott, *The Message of Genesis: A Theological Interpretation* (Nashville: Broadman Press, 1961), 39,40.

life spans of 110 years were symbolically attributed to those leaders who had lived full, complete lives.[77]

Bernard Ramm, the author of a leading text on hermeneutics (Bible interpretation), suggests that the ages might have represented the life spans of dynasties named for their leaders. Adam's dynasty, in this view, would have existed for about a century before Seth, one of his great-grandsons, split off and started another community ruled by his ancestral line. Adam's dynasty continued to exist for another seven hundred years after the divergence.[78]

Whereas Hill and Ramm primarily seek literary explanations for the long lives of the patriarchs, astronomer and apologist Hugh Ross explores the possibility that the patriarchs really did have extremely long life spans. He notes, in particular, the shift from a vegetarian diet to a diet that included meat in the years following the Adam's flood. Was this because the flood had destroyed most of the edible plants, or was something else going on? The eating of meat causes a gradual buildup of heavy elements in the bodies of those who eat it. For beings with life spans of less than one hundred and twenty years, this is not a problem. If human life spans were somehow extended beyond this limit, however, the accumulation of these elements would eventually lead to health problems.[79]

Naturally, many of those outside the Christian community consider the advanced ages of the patriarchs to be strictly legendary in nature, but even these explanations are incomplete if they do not address the question of why, even in legend, the

---

[77] Carol A. Hill, "Making Sense of the Numbers of Genesis," *Perspectives on Science and Christian Faith*. Vol. 44, No. 4, Dec 2003, pp. 239-251.

[78] Ramm, *Science and Scripture*, 236-237.

[79] Hugh Ross, *The Genesis Question* (Colorado Springs: Navpress, 2001), 119.

authors would have attributed such longevity to the characters. So far, studies of ancient human remains have not uncovered any evidence that ancient humans lived any longer than the people of today. What kinds of evidence would scientists look for if they had? Would the telomeres (which limit the number of times a cell can reproduce) in the cells of the people who lived before Noah's time have been ten times longer, or would they somehow be absent altogether or replaced by something else? Or, would God have supernaturally lengthened people's lives, intervening directly without making any biological change in them? Would the buildup of heavy metals be strangely elevated in the cells of some?

Interestingly enough, the Hebrews were not the only Near Eastern civilization of that era to write about the long life spans of ancient people. The Sumerian King List (a Sumerian text dated to about the end of the third millennium B.C.) provides a listing of kings who ruled before and after the flood. Those before the flood lived to be thousands of years old. After the flood, they shortened to more modest lengths.[80]

Does this indicate a biological change, as Ross suggests, or is it, as Hill suggests, a matter of numbers having a religious significance that is completely lost on modern readers? If numerological age and divinity, in the minds of the ancients, were somehow linked, the divinity numbers would have been removed from the ages of those who lived after the flood. If the civilizations after the flood were seen as pale imitations of the golden age when God (or gods, in the Sumerian version) "strived with men," their "divinity scores" would have been much lower.

---

[80] Clyde T. Francisco, "Genesis," in *The Broadman Bible Commentary, Vol. 1 General Articles, Genesis, Exodus.* Clifton J. Allen, General Editor. (Nashville: Broadman Press, 1973), 135, 136.

CHAPTER 5:
A Guided Tour of Genesis, Part III: Enter the Nephilim

The sixth chapter of Genesis has always fascinated me. It tells the story of a mysterious group of men referred to only as "Sons of God." These men marry the "Daughters of Men" and give birth to the heroes of legend. The *Nephilim,* or "fallen ones," are said to have dwelled on the earth in those days (Genesis 6: 1-4). Scholars have interpreted these verses in a variety of ways. Before we try to tackle the historical issues surrounding these verses, I'd like to call attention to their religious purpose.

As I noted at the beginning of the section, the main challenge the authors of Genesis faced was not science-based skepticism but the pagan belief systems of neighbors like the Canaanites, the Egyptians, and the Babylonians. Just as Greek and Norse mythology had their stories of men like Hercules and Thor, who were produced by sexual unions between human women and gods, the civilizations around Israel had "god-men" as well.

The question for the writers of Genesis was not how something like that could have occurred biologically, but where beings like these, if they existed, would fit into a monotheistic (single God) religion. Genesis portrays them not as gods to be worshipped but as intruders who corrupted God's creation. Just as the fourth-day creation of the sun, moon, and stars puts the celestial bodies in their places not as gods that mystically determine the fate of men and women but as created things, this chapter seems to be putting the gods and demigods of the surrounding cultures in their places as well. It is, admittedly, a strange set of verses. As a fan of fantasy worlds like the one shown in *The Lord of the Rings*, I enjoy the passage because of the mysterious ancient world portrayed there, and I enjoy fantasizing about what it would be like to unearth the buried remains of such a world. As a would-be Bible scholar trying to explain the passage to skeptics and avoid alienating other Christians, I find the passage raises some thorny questions.

One of the most widely accepted interpretations of this passage is that the "Sons of God" were fallen angels who somehow found a way to breed with human women.[81] If the goal of Genesis was to establish the Creator-God of the Hebrew religion as the ruler over the "gods" of the surrounding cultures, this interpretation makes a great deal of sense. It is also supported by Jewish tradition. *The Book of the Secrets of Enoch*, a piece of Jewish literature written sometime between 200 B.C. and the early first century, is fictitiously written from the point of view of Enoch, the father of Methuselah, who lived in the years before Noah's flood. The author refers to these sons of God as "the Watchers" and says they were demonic beings, fallen angels who had violated their place in the natural order by having sexual relationships with human women.[82]

Whereas some scholars view the "sons of God" passage as a literary accommodation to the widespread beliefs of the people of that time, many conservative scholars are not comfortable with this mixing of history and legend so they look for a more concrete way to understand the passage. Treating the passage as literal history presents problems of another kind. If these

---

[81] William S. Lasor, David A. Hubbard, and Frederic W. Bush, *Old Testament Survey* (Grand Rapids: Eerdmans Publishing Co, 1996), 27.

[82] Willis Barstone, *The Other Bible* (New York: HarperOne, 2005), 487-488.

"sons of God" were demons, how could spiritual beings have reproduced with flesh-and-blood women?

One possible solution is to view the passage as an attack on pagan fertility cults that involved temple prostitution rites in which gods symbolically mated with human hosts.[83] In later Mesopotamian civilizations, most notably Babylon, the idea of priestesses participating in marriage rites to the gods was a common practice. Herodotus, a Greek historian who visited Babylon in 460 BC, wrote of a temple at the summit of the ziggurat[84] there. A Babylonian woman, chosen by the god Marduk, was supposed to spend the night alone in the temple in a symbolic marriage ceremony as part of the *akitu* New Year's ceremony.[85]

Whether supernatural beings were really involved at all, or whether the people of that time only thought they were, the children of the women who participated in these rites might have been considered to be proxy children of the gods. Either way, a connection of the events described in Genesis to the pagan temple practices that were common in both Mesopotamia and Canaan[86] is one possible way of accommodating both pagan beliefs and literal history.

The interbreeding of human women with demonic entities is not the only explanation for the passage. Another explanation is that these sons of God were powerful rulers. Since many

---

[83] Gordon J. Wenham, *Word Biblical Commentary, Volume 1, Genesis 1-15* (Nashville: Word Publishing, 1987), 141.

[84] A ziggurat is a pyramid-like structure in which each level is narrower than the one beneath. The ziggurats in the ancient Near East were made of sun-dried brick.

[85] Henri-Paul Eydoux, "The Men Who Built the Tower of Babel," in *The World's Last Mysteries* (Pleasantville, NY: The Reader's Digest Association, 1978), 177-178.

[86] Alfred J. Hoerth, *Archaeology and the Old Testament* (Grand Rapids: Baker Academic: 1998), 66-70; 119-222.

rulers of that time claimed to be divine sons of the gods, the distinction between gods and men would have been blurred in the minds of the ancients. "Sons of god" would have been a religion-based caste designation. The higher-caste ruling class would be viewed as coming directly from God, while the lower castes would be considered to be little better than cattle. (Gilgamesh, for example, was viewed in Akkadian literature as a historical figure, thought to be two-thirds divine and one-third human.[87]) This explanation is completely plausible.[88] Another solution is that the sons of God represented the righteous line of Seth, whereas the sons of men refers to the corrupt line of Cain. Even the children of the more righteous son, by the time of the flood, had become corrupt.[89]

The sons of God passage is one of the most interesting and the most confusing found in the book of Genesis. Theologically, it is a way of explaining the myths of demigods found in the cultures that surrounded Israel. The historical roots of the story are more difficult to determine. Regardless of one's take on this issue, however, the results are the same. Civilization had become hopelessly corrupt, and God's heart was grieved. His efforts to redeem corrupt society had met with only corruption and indifference, and it was time to act.

---

[87] Hamilton, *The Book of Genesis*, 264.

[88] Ibid.

[89] Ibid., 265.

CHAPTER 6:
A Guided Tour of Genesis, Part IV: God Hits the Reset Button

Whatever meaning theologians may assign to the strange episode of the Sons of God and the mysterious Nephilim, early civilization has become so corrupt that God decides to wipe the slate clean and start over by means of a catastrophic flood. Because creation, in Genesis 1, is shown to have emerged from a state of watery chaos, the original readers would have viewed destruction by water as God's way of returning the created world to its original disordered state. God was, in effect, hitting the reset button.

One family, however, is chosen to survive and repopulate the earth. This is the family of Noah. God instructs Noah to build an ark and to fill it with breeding pairs of the various animals found on the earth. Noah does as God instructs, the flood sweeps the earth, and only Noah and his family survive (Genesis 6:9-22).

The Noah story presents a variety of plausability problems for modern readers: Where did all of the water come from? Could all of those animals really have fit into a single structure? How did they get from distant places like Australia and South America to Mesopotamia? What happens when you mix freshwater and saltwater fish? Faced with questions such as these as well as the discoveries of geologists since the 1800s, Christian scholars have debated whether or not the narrative requires a planetary catastrophe or whether Noah's flood could plausibly be conceived of as a local event that only appeared to be universal from the perspective of the author. His world, some scholars say, need not have been the entire planet. Is there any biblical reason for such an interpretation, some

Christian and secular readers might ask, or are local flood theories just a cop-out for evangelical intellectuals who want to have their cake and eat it too? Let us examine the arguments.

In Genesis 8:21-22, God makes a covenant with his creation that he will never again destroy every living thing as he has done in this instance. As long as the earth remains, God says, the days and nights and the seasons of the year will remain the same. This promise is given in the form of verse. These passages, if read as literal history, make the strongest Bible-based argument that the account was meant bo be read as a global catastrophe.

It is important to remember, however, that neither the prophet nor his audience would have had any concept of earth as a planet rotating in space. According Whorton and Roberts, the phrase *kol erets*, translated "entire earth," appears 205 times in the Old Testament and at least 80% of these references clearly refer to a local region. In Genesis 41:56-57, for example, a famine covers the whole face of the earth. In Genesis 13:9-10, Lot tells his nephew the whole land/earth is before him. In Leviticus 25:9, on the year of the Jubilee, the children of Israel were commanded to make the trumpet sound throughout the whole land/earth.

Also see Zephaniah 1:2-3, where it sounds as if God intends to destroy the whole world for the sins of Judah. (The word for land in Zephaniah, however, is *adamah,* rather than *erets*.) The phrase *kol shamayim*, "entire heavens" sounds even more conclusive, but it is also used to refer to a limited area. Deuteronomy 2:25 is an example. In it, God tells the wandering Israelites he will put the fear of them upon the nations under heaven.[90] According to many Bible scholars, insisting that the flood must be global or the Bible cannot be true places unnecessary demands on the text.

---

[90] Mark Whorton and Hill Roberts, *Holman QuickSource Guide to Understanding Creation* (Nashville: Broadman and Holman, 2004), 157-158.

Although the geological discoveries of the 1800s may have challenged the traditional understanding of the text, they did not automatically discount the possibility of a historical flood event, even for people who do not consider the Bible to be divinely inspired. The exact time, extent, and location of the event remain matters of speculation.

In 1929, archaeologist Leonard Wooley made the startling announcement that he had "dug up the flood" while excavating the site of the ancient city of Ur. Between two occupied strata, Wooley's team found a twelve-foot-thick layer of alluvial mud obviously deposited by a massive river flood. Later expeditions, digging in nearby locations, failed to find evidence that Wooley's flood was anything but a local phenomenon. Wooley continued to believe, however, that this was the historical root of the biblical flood story and the various flood legends found throughout the region.[91] Other authors link the flood narrative to the catastrophe that formed the Black Sea around 8,000 BC.[92] There is no consensus among scholars, Christian or otherwise, about whether either of these flood events was *the* flood and speculation abounds.

The story also indicates that there was no rainfall on the ground in the area of Eden before the flood (Genesis 2:5). Though some believe this indicates a massive re-ordering of the laws of nature, others believe the narrative may refer specifically to the lack of rain in southern Mesopotamia. There may be literary and theological issues involved as well. Gary Rendsburg, a Jewish scholar, says that the Hebrews of the era in which Genesis was written sometimes viewed water as a symbol for evil. Note the lack of the phrase "God saw that it was good" in Genesis 1:7, when God orders the oceans and waters above the sky.[93] The significance of rain in Elijah's

---

[91] Werner Keller, *The Bible as History* (New York: Bantam, 1981), 22-30.

[92] Ian Wilson, *Before the Flood* (New York: St. Martin's Press, 2001).

[93] Gary A. Rendsburg, "Lecture 2: Genesis 1," *The Book of Genesis.* The

showdown with the prophets of Baal (1 Kings 17:20-46) takes on new meaning when one realizes that Baal was the Canaanite rain god. By holding back the rain, God, in essence, humbled Baal.

The original audience, a people subject to violent annual floods, would not have been concerned about the debates over geological evidence. As I pointed out earlier, they would not have conceived of the earth as a planet orbiting a star or even had the vocabulary to describe such a concept. The most important idea was the idea of a God who made a promise to them. They didn't understand meteorology or ice ages or where water comes from, and so far as they knew, the whole world could sink beneath the ocean at any time, ...*but God had made a promise.* In an age when we worry about asteroid strikes and climate change, perhaps the idea of God promising to take care of his people doesn't seem so strange after all.

In the years following the flood, the children of Noah rebuilt civilization and began work on a great tower that would reach to heaven. All of them were gathered in one place and speaking a single language. God, displeased with this state of affairs, stopped the tower's construction by "confusing the languages" and scattering the children of Noah over the face of the earth (Genesis 11:1-9). This brief story leaves many questions. According to the text, in Genesis 11:6, God saw that "nothing they purpose to do will be impossible for them." Why was this growth in capabilities considered to be a bad thing? Why did God find it necessary to scatter the people? The primary reason given in commentaries is that they defied God's command to fill the earth and gathered in one place to build a monument to their own glory.[94] Was this "tower to heaven" a seat of pagan

---

Great Courses audio lecture series, produced by The Teaching Company of Chantilly, VA. 2006.

[94] Victor Hamilton, *The Book of Genesis, Chapters 1-17,* vol. 1 of *The New International Commentary of the Old Testament* (Grand Rapids: William B. Eerdmans Publishing Company, 1990), 353.

worship as well? The records of extrabiblical historians lend possible support to this notion. The tower of Etemenanki, located in the heart of Babylon, was a center of pagan worship in the time of Herodotus (mentioned previously) and is considered by some to be the historical Tower of Babel.[95] Francisco, however, believes the text refers to an earlier tower from a time in the more distant past. The text, he says, was written at a time when the ziggurat in Babylon was in ruins. He believes the people of Babylon built that particular pyramid in an attempt to rebuild the tower described in the *Enuma Elish*, a Babylonian creation story.[96]

Readers may find some irony in the way God has to go down to reach the "tower to heaven," but God does not laugh at the people's futile attempt to storm heaven's gates. He says, on the contrary, that they will soon be able to do whatever they want and puts a stop to their plans.

Following the lead of the Genesis author, Jewish and Christian writers have linked the word translated "Babel" to the Hebrew word *balel*, which means "confusion." The people of the region, however, say that it came from the word *Bab-ili*, which meant "Gate of the Gods." Commentary writers believe the writer of Genesis was writing in deliberate mockery of the mighty works of the Babylonians. What they considered to be a mighty gate to the gods was really only a confused mess.[97]

The text is not completely clear on how God confused the languages. Is the story a fable, as some assert, explaining why

---

[95] Henri-Paul Eydoux, "The Men Who Built the Tower of Babel," in *The World's Last Mysteries* (Pleasantville, NY: The Reader's Digest Association, 1978), 169-178.

[96] Clyde T. Francisco, "Genesis," in *The Broadman Bible Commentary, Vol. 1 General Articles, Genesis, Exodus.* Clifton J. Allen, General Editor. (Nashville: Broadman Press, 1973), 151.

[97] Bruce K. Waltke, *Genesis: A Commentary* (Grand Rapids: Zondervan, 2001), 178.

people don't all speak the same language? Gordon Wenham, in the *Word Biblical Commentary*, views the episode as a rebuttal of a pagan myth of that time. Though the story of the tower has no exact parallel in Near Eastern literature, there is a story about the confusion of languages. The god Enlil wanted everyone in the world to speak the same language so all could worship him. Enki, a rival god, thwarted his plan by confusing the languages. In studying Near Eastern literature, scholars have noted that the creation stories of the people who surrounded the Hebrews taught that events like the ordering of the world and the flood were the result of battles between gods.[98] The challenge of Genesis was to explain the chaotic as well as orderly elements of creation with only one mighty God. The conflict, in the Hebrew religion, was not between gods but between God's goodness and humanity's stubborn disobedience. Satan and his demonic forces did not play the major role in the earlier Old Testament books that they did in later books like Daniel.

Many conservative Christians are troubled, however, by the suggestion that the Babel story does not represent a literal historical event but a literary response to pagan beliefs. They view the story not as sanctified mythology but as restored history. Viewing the story from that perspective, we might wonder whether the sudden inability of the people to understand spoken language might have been caused by a medical condition like aphasia that affects the language center of the brain. If God caused one generation to be struck with a condition that rendered them unable to communicate, civilization would certainly have been stopped in its tracks. The confusion might also be explained by a direct and localized intervention in which God temporarily scrambled the perceptions of the builders. The biblical book of Acts describes a miracle in which God allows people from around the world to hear their own languages spoken by the apostles on the day of

---

[98] Gordon J. Wenham, *Word Biblical Commentary, Volume 1, Genesis 1-15* (Nashville: Word Publishing, 1987), 235-237.

Pentecost (Acts 2:1-8). "And when this sound occurred," the passage says, "the multitude came together, and were bewildered, because they were each one hearing them speak in his own language." Was it the apostles' speech that God altered or, as some have suggested, the perceptions of the hearers?

The genealogies, for many of us, are the least interesting parts of the Bible, and many of us tend to skip over them when we read because of their lack of devotional value. The Table of Nations found in Genesis 10 and 11, however, merits special attention. According to Bible scholars, the table is unique (as far as they know) among the documents of that period. Its sweep of the known world of that time goes as far north as the Black Sea, as far south as the source of the Nile, as far east as Spain, and as far west as the Iranian plateau. No other known document from that time period attempts to describe so many people groups.[99] Most scholars do not believe the table contains all of the nations known to the author(s), only those of importance to the Israelites at the time Genesis was written. There is evidence of stylization in the table's arrangement. The number seven, which symbolizes completeness, occurs frequently in the document, with the total number of nations adding up to 70. (The list of Jacob's family group, given later in Genesis, also adds up to 70.)[100]

Even if some nations were excluded, it would have been unusual for the average person of that era to know of all of the nations included in the table. This leads some to believe the author was from one of the large city-states of that day, possibly a member of the royal court of Egypt or Jerusalem. Many scholars date the completed table to the first millennium B.C., because some of the groups mentioned do not appear in material outside of the Bible until that time.[101] The book of

---

[99] Francisco, "Genesis," 149.

[100] Ibid., 149-150, and Wenham, *Word Biblical Commentary*, Volume 1, 214.

[101] Wenham, *Word Biblical Commentary*, Volume 1, 214.

Ezekiel, written in that time period, mentions many of the same nations (e.g. Magog, Meshech, Tubal, Cush, Put) in chapters 27, 28, 32, 38, and 39. If the table was miraculously revealed, of course, dating it in this way would not necessarily be considered valid, and the presence of material unknown to most of the world would be evidence of its divine origin. Miraculous inspiration, however, does not yield itself to scholarly study.

Scholars also debate whether the listing is purely a listing of ancestors, like a modern genealogy, or whether the writers had other goals in mind. The children of Japheth, given first, were seafaring nations that had little direct contact with the Hebrews.[102] The sons of Ham were city-dwelling nations[103]. Many of them, including the Egyptians, Babylonians, and Philistines, had been enemies of Israel.[104] The sons of Shem, Israel included, were mostly nomadic peoples. They were the good guys of the story, so to speak, and were considered to be the chosen line.[105] Note that here, as in the case of Cain and Seth, that the corrupt dwell in cities, and the faithful are nomadic farmers.

This distinction speaks to prejudices that persist to our own time. If a timeless, transcendent God were to select a group of people to reveal himself to, why would he select a group of wandering nomads? Wouldn't it make more sense for God to reveal himself to an advanced civilization like the Sumerians, the Egyptians, or the Babylonians? Great cities with towering monuments, the author of Genesis tells his generation and ours, tell us more about human pride than spiritual greatness. The

---

[102] Ibid.

[103] Ibid.

[104] Waltke, *Genesis: A Commentary*, 163.

[105] Wenham, 214.

Christian New Testament builds on this theme. In an era in which the military might of Rome and the wisdom of the Greeks dominated western civilization, God chose to reveal himself to the world through a carpenter from a tiny country rather than through a Roman Caesar or a Greek philosopher.

Following the Table of Nations and the confusion of languages at Babel, the perspective of Genesis shifts radically from a cosmic scale to a local one. It focuses in on a man named Abram (later renamed Abraham) living in the land of Ur and on the role of his family in the birth of the Hebrew nation and the worship of the one God. The God of the Hebrew nation would later be referred to as "the God of Abraham, Isaac, and Jacob."

The number of generations from the children of Noah to the time of Abram appears short in the narrative. If the genealogy is taken at face value, with no allowance for gaps, it appears that Noah was still alive at the same time as Abraham, and that his son Shem outlived Abraham.[106] Archaeology (as well as the content of the story) suggests a much longer time span between Noah's flood and the founding of the civilization from which Abraham emerged.[107] This, among other things, leads many Bible scholars to view the genealogies as outlines rather than exhaustive lists. If there are gaps, however, they appear to have been between clusters of related characters rather than occurring at even intervals between all characters (i.e. If a father and son appear in the same narrative, they apparently lived at the same time).

---

[106] Carol A. Hill, "Making Sense of the Numbers of Genesis," *Perspectives on Science and Christian Faith*. Vol. 44, No. 4, Dec 2003, 244.

[107] Alfred J. Hoerth, *Archaeology and the Old Testament* (Grand Rapids: Baker Academic: 1998), 31-74.

CHAPTER 7:
Genesis and Archaeology

As noted, many Christians through the centuries have taken the events described in the early chapters of Genesis at face value and viewed them as literal historic facts. The discoveries of the past two centuries, however, have led scholars to paint a longer and more complicated picture of human origins and development. Before discussing the scientific "timeline," however, it is instructive to reflect upon the time and place from which Genesis originated.

The books of the Bible did not emerge from a cultural vacuum. Though the messages of Scripture are timeless, the books themselves were written in specific times and places for specific people. To understand a book of the Bible fully, it is helpful to understand its context. The book of Genesis presents some challenges, because the exact date and nature of its composition are unknown. Although some books of the Bible tell of prophets receiving messages directly from God, most are silent about the details of the composition process.

Tradition holds that Moses composed the book of Genesis, along with the books of Exodus, Leviticus, Numbers, and Deuteronomy. Modern critical scholarship paints a less straightforward portrait of the process. According to the JEDP theory, the five alleged books of Moses were actually composed over a period of centuries by different authors in different places. Even if one does not hold to this theory in its entirety (see Duane Garrett, *Rethinking Genesis*,[108] for an interesting and scholarly alternative), there is evidence in the

---

[108] Duane Garrett, *Rethinking Genesis* (Grand Rapids: Baker, 1991).

text itself that its composition was not as simple as tradition would lead one to believe.

The book of Genesis is a special situation regardless. Unlike the events described in the other four books ascribed to Moses, the events described in Genesis took place well before the time of Moses. They were almost certainly preserved as oral traditions or, perhaps, as fragmentary written documents. They were part of the culture and identity of the Hebrew people.[109] Viewed this way, Genesis might well be viewed as "a Bible within the Bible" or as Moses' Old Testament.

The discoveries in Ninevah in the late 1800s shed light on the belief systems of the Sumerians, Babylonians, and other Mesopotamian groups, as ancient myths like the *Epic of Gilgamesh*, the *Enuma Elish* (meaning "when on high"), and the *Epic of Atra-Khasis* (meaning "ultra wise") came to light. Though expressed in terms of gods, goddesses, and monsters, the stories provided a composite picture of the way the people in that time and place viewed the universe. Like the creation week narrative in Genesis, the world began with watery chaos. Genesis calls it *tehom*, the deep. The *Enuma Elish* gives it the identity of Tiamat, a female sea monster who is a living personification of the saltwater sea. The land, in the old cosmology, sits above the deep. A bowl-shaped firmament spread across the sky shields the earth from a great freshwater ocean, referred to by scholars as the *supercealean sea*, that sits above the earth. In the *Enuma Elish*, Marduk the sun god kills Tiamat, splits her apart like a shellfish, and uses her upper shell to form the firmament that keeps the fresh water sea in its place above the sky and her lower shell to form the land.[110]

Readers interpreting Genesis are faced with a question of how to read the details of the universe they find described

---

[109] William S. Lasor, David A. Hubbard, and Frederic W. Bush, *Old Testament Survey* (Grand Rapids: Eerdmans Publishing Co, 1996), 6-14.

[110] Morris Jastrow, Jr., "The Hebrew and Babylonian Accounts of Creation." *The Jewish Quarterly Review* 13 (July 1901): 626-630.

there. Would a writer, even a divinely-inspired one, write his narrative using a model of the world as people understood it at that period of time, or would God have given him superhuman knowledge of the universe?

Some modern Christians have tried to mine the narrative for scientific data and have used details like the firmament and the "waters above the waters" to develop some unusual views of the universe. One theory holds that the early earth was surrounded by a suspended canopy of water that collapsed at the time of Noah's flood. Many Christian scholars believe that these Christians are failing to account for the original author's understanding of the world, that this was simply the way the people of that era understood clouds and rain (Psalm 104, in verses 3 and 13, mentions the firmament as a present-day reality), and that one should simply make the adjustment.[111]

What is most unique about the book of Genesis is not its cosmology but its theology, specifically its views of God and of humanity. Though supernatural beings appear frequently in Mesopotamian mythology, the idea of a single, transcendent God who created the entire physical universe is original, to say the least. Many of the gods of the surrounding cultures were personifications of natural elements like the sun and the sea.[112] The exalted place of humans in the created order is quite different from the lowly place they held in the Mesopotamian myths, where they were created primarily to serve as laborers to grow food for the gods. Although one might discern some echo of this in the Genesis statement that "there was no one to work the ground" (Genesis 2:5), the idea of humans being made in the image of a transcendent, eternal God is as completely foreign as the idea of such a God existing.

---

[111] Bernard Ramm, *The Christian View of Science and Scripture* (Grand Rapids: William B. Eerdmans Publishing Co., 1954), 159, 160; and Newport, 154.

[112] John Newport, *Life's Ultimate Questions*. (Dallas: Word Publishing, 1989), 133.

What conclusions, if any, can we draw from the study of the biblical creation account in relation to other Near Eastern accounts? Without a library of documents containing earlier versions of the Genesis creation accounts, theories about the details of the book's composition are difficult to prove and are all too vulnerable to the preconceptions of the various investigators.

We may draw two conclusions with relative certainty, however. One is that the Bible was not composed in isolation from the world that surrounded it. The imagery found within its pages is that of the ancient Near East. Rather than discounting the backgrounds and histories of the Bible's human authors, Christian readers should say instead that this is the time and place God chose to begin his story. The Mesopotamian roots of Genesis may be seen as part of his plan.

What can we learn about the beliefs of the ancient Mesopotamians by reading their myths? We learn that they believed the world began with watery chaos and rain came from a freshwater ocean above a barrier in the sky. They also believed in a catastrophic flood that nearly wiped humanity as they knew it out of existence, and in a good and wise man favored by the gods, who was spared from the disaster. They also apparently believed that people who lived before the flood had longer lives than those who lived afterward. The writers of the Bible seem to have shared these beliefs.

The second conclusion is that the moral and transcendent God portrayed within its pages stands alone among the often petty and all-too-human gods of the other Near Eastern cultures. The God of Abraham, Isaac, and Jacob is not the product of Babylonian, Assyrian, or Sumerian theology. Neither is he Baal, or the rain god of the Canaanites. Egyptian sun worship offered a monotheism of sorts, but the God of the Hebrews created the sun and was before all physical things. Ahura Mazda, the "Wise Lord" of Zoroastrianism, a Persian religion, may be closer to the Hebrew idea of an all-wise creator God than any of the others of that day. Compared to most, though, one can truly say that the most remarkable feature of the Hebrew creation account is the nature and personality of God himself.

So, what options do these discoveries leave Christians with when it comes to describing the Genesis narratives? We are not asking, at this point, whether the stories were inspired by God or free from error. We're asking instead what kind of literature they are. Two primary ways of viewing them might be called the "Sanctified Mythology View" and the "Restored History View."

The Sanctified Mythology View assumes the writer (or writers) of Genesis took imagery that permeated the cultures of the ancient Near East and wove it into an epic that honored his God and set him apart from the gods of the surrounding cultures. The Restored History View would hold that the events in Genesis were actual historical events that were remembered "through a glass darkly" by the pagan cultures of Mesopotamia and that when God inspired the book's composition, he set the record straight for the author and his readers.

The Sanctified Mythology View does not assume the Hebrews just copied the mythologies of the surrounding groups because they could not think of their own. Of the stories in Genesis, Noah's ark is the only one that has a clear narrative parallel. The other similarities (e.g. Gilgamesh's episode with the fruit and snake) are more fragmentary.

Genesis is not, according to this view, a bad copy of Akkadian mythology but a completely different way of looking at the world. While the people around the Hebrews worshipped the sun, moon, stars, and sea and believed in human-like gods that ate and slept and had sexual relations with human women, the God of Genesis was invisible and transcendent. Instead of being just another created thing, this supremely powerful God existed outside of Creation and shaped it, formed it, and governed it. Though that may seem so obvious to modern readers that it barely deserves comment, the Genesis model was a major departure from the way the surrounding cultures viewed the universe.

In the Sanctified Mythology View, Genesis answered the question "How is your God different from all of the other gods?" by placing God in a sweeping epic that paralleled the epics of the surrounding cultures in many ways but was startlingly different in others. To hold this view is not necessarily to reject God as the book's spiritual author but to change one's idea about how the process of inspiration took place.

One question the Sanctified Mythology View does not answer for us is where the Hebrews got this new theology in the first place. What literal, historical events had they witnessed that led them to believe in this invisible Creator-God? Were they convinced by Abraham's encounters with God in visions, the experiences of Moses and the Israelites surrounding the Exodus, or something else? Those who do not believe in the supernatural would explain the development purely as an evolution in thought processes. Perhaps the Hebrews outgrew the nature gods and wanted a more sensible theology. For many of us, however, that answer is just not satisfying. It is like theorizing that the early church simply invented Jesus.

The Restored History View is more appealing to many American Christians for a number of reasons. The first is one of familiarity. For many of us, it is simply the way we have always been taught. Another is that we don't use mythology in the same way those ancient cultures did and don't really understand the important role it played for them. For us, mythology is just something we use to entertain children or something ignorant people believed in because they didn't know any better. The word "myth" also has a negative connotation. We consider it a synonym for "lie" or "misconception." In anthropology, however, a myth is a story that embodies the values of a group of people. The story of young George Washington chopping down a cherry tree and telling his father the truth about it is an American example of a myth. It may or may not be historical, but it uses a leading American figure to teach the value of honesty. For some of us, the idea of those stories being less than completely literal amounts to doubting the power of God or the Resurrection of Christ.

Some readers would object that God would never have allowed his Word to be tainted by references to pagan mythology, but the Bible's authors did apparently make some use of stories found in the surrounding cultures. Psalms 74:13-14, Isaiah 57:9-10, and Job 26:12-13, for example, describe God's battles with great sea monsters (Rahab and leviathan) described in Canaanite, Babylonian, and Hittite mythology.[113] The book of Jude, verse 9, refers to a confrontation between Satan and Michael the archangel over the body of Moses. The confrontation he refers to does not come from the Old Testament but from a piece of extrabiblical literature called *The Assumption of Moses*. He also refers to the book of Enoch, another piece of extrabiblical literature, in verse 14.[114] Do these references force us to argue for the literal historicity of Rahab, leviathan, and the confrontation Jude refers to because they appear in the Bible, or should we simply view them (some of them, at least) as sermon illustrations? When a pastor refers to Cinderella, Frankenstein, or Darth Vader from his pulpit, he is not lying to his congregation or using the authority of his position to argue for a historical Cinderella, Frankenstein, or Darth Vader. Cinderella might be used to represent grace. Frankenstein and Darth Vader both represent pride leading to disaster or good intentions gone awry. It is the potency of these characters as cultural symbols, not their historicity, that makes them such good illustrations. This does not mean, of course, that we should dismiss all of the Bible's miracle stories as mythical or metaphorical, but we should pay attention to the

---

[113] Conrad Hyers, *The Meaning of Creation* (Atlanta: John Knox Press, 1984) 62.

[114] David Dockery, General Ed., *The Holman Concise Bible Commentary* (Nashville: Broadman and Holman, 1998) 655-656.

literary form the author is using in the given verse. In some cases, however, our unfamiliarity with the literary forms of the time periods in question makes this difficult to determine. Poetry and prophetic visions are particularly difficult to understand. Imagine trying to make sense of a political cartoon 3,000 years from now with no knowledge of the political situation it is referring to. If you do not know Uncle Sam represents the U.S., donkeys and elephants represent Democrats and Republicans, respectively, and that a bear can represent either Russia or an economic market situation (bear market or bull market), you'd have a very hard time making sense of it.

This brings us to the question of how the New Testament authors use the Old Testament stories. Characters from Genesis show up in the genealogies of Christ. Many consider St. Paul's references to Christ as the "second Adam" to be an affirmation of a historical Adam and Peter's mention of Noah's flood to be an affirmation of a literal, historical Noah.

Some of the images in the stories, however, still look more like the kind of symbolic imagery we would find in mythology or in apocalyptic books of the Bible. The imagery in the Eden narrative, especially, more closely resembles the prophetic imagery in the book of Revelation than the historical narrative found in the gospels. That brings us to the third option. It may be that the stories are built around a literal, historical core but are expressed in a way that is less precise but more colorful (like the political cartoons I mentioned earlier) than what we would find in modern scientific and historical writing. Knowing exactly where to draw the line between the two is tricky, however, because some elements of the stories (e.g. the description of the location of Eden) read more like history whereas other elements (e.g. the serpent, the fruit) look more like literary symbols.

Again, this is not a question of how "high" one's view of Scripture is. It is not a question of whether God inspired Genesis or whether it is a trustworthy source of information about him. Those questions will have to wait for another volume. The question is one of literary style and just what kinds of information it was supposed to convey. Clearly, the purpose was to answer the question, "How is our God different?" and Genesis answers that question regardless of which of the two views you hold. The question is whether the author(s) answered that question using the kind of precise scientific and historical language Westerners are used to, or whether it used the more colorful and less precise language that ancient Near Eastern people used when they talked about gods.

In the end, those who believe that Genesis is "just another Near Eastern myth" are still left to wonder why Genesis and the religions it helped to spawn have managed to survive for many centuries when the myths and civilizations of larger and more powerful neighbors were buried and forgotten. Was it just the luck of the draw or was it something more? Christians and Jews would say it is the living God behind the stories that has kept them alive for so many centuries. Those who do not believe in God, of course, look for more mundane explanations.

***

In the chapters that follow we will explore the creation narrative put forth by cutting-edge science and wrestle with the question of how a narrative assembled from telescope images, geological studies, fossils, and archaeological discoveries might relate to a religious document assembled between 2,400 and 3,400 years ago (depending on whether one believes Genesis was assembled by Moses in the thirteenth or fifteenth century B.C. or by an unknown compiler during the Babylonian exile between 500 and 600 B.C.). Are belief in science and belief in the Bible mutually exclusive options, or are there ways to reconcile the two without doing violence to either one? Are we better off not trying to resolve the issues and simply appreciating each story on its own terms? Some readers might prefer that option. Even if we cannot resolve all of the issues, we can at least examine some of the ways the thinkers of the past have attempted to deal with them and perhaps further our own understanding.

CHAPTER 8:
A Guided Tour of the Scientific Timeline, Part I: The Big Bang

Now we leave behind the first museum with its ancient scrolls, archeological dig sites, and date palms along the Euphrates and step into the second. After passing through an atrium filled with dinosaur skeletons, we step into a domed theater, climb the stairs, and settle into our padded seats. The room goes dark; a tiny spot of light appears on the screen, and a massive explosion flashes silently into nothing as the seats vibrate beneath us. The universe, once again, has begun.

Space and time, according to the currently accepted scientific theory, began with an explosion, a "Big Bang" some 13.7 billion years ago. This theory is the culmination of a series of discoveries. One of the first and most significant was Edwin Hubble's discovery, in 1929, that there were other galaxies besides our own, and they were moving away from each other as though they had once been closer together.[115] By "reversing the arrow" of their movements, later researchers were able to estimate the time of the explosion that had hurled them outward. The development of more and more powerful telescopes allowed them to see further into the past, so to speak, by picking up light that had been traveling across the universe almost since its beginning. The Big Bang Theory gained momentum as new discoveries were made. It competed for a while with the Steady State Theory, a theory that

---

[115] On a technical note, some galaxies, like our nearest neighbor Andromeda, are actually moving closer. They move in clusters rather than independently of each other.

proposed the universe had no beginning or end, and ultimately triumphed over it. [116]

Though the exploding universe model is the most widely accepted model of the way our universe formed, the cause of that primeval explosion is unknown. In this theory, space is envisioned as the surface of a balloon that is inflating into nothingness and expanding as it goes.[117] (A more in-depth description of the Big Bang, as modern science conceives it, would require an understanding of quantum mechanics that most of us do not have.) Some Christians see the Big Bang Theory as an attack on faith by scientists. Others are enthusiastic about the Big Bang because it is consistent with the idea of a universe that has an identifiable beginning. The reactions of some leading scientists to discoveries supporting Big Bang cosmology are worth noting.

In order for the universe to be able to support life by forming heavy elements, stars, and planetary systems, many forces had to be properly balanced. Included in this list are factors having to do with the force of gravity, the speed of light, the strong nuclear force, the weak nuclear force, and electromagnetism. The estimated odds against these forces being balanced as they are has produced reactions of wonder from scientists who have described this accumulation of phenomena as the Anthropic (meaning *manlike*) Principle.[118]

Physicist Freeman Dyson, after going over the convergence of events, concluded, "The more I examine the universe and the details of its architecture, the more evidence I find that the universe in some sense must have known we were coming."[119]

---

[116] Francis S. Collins, *The Language of God* (New York: Free Press, 2006), 64.

[117] Andrew R. Liddle, Andrew R. and David H. Lyth, *Cosmological Inflation and Large-Scale Structure* (Cambridge: Cambridge University Press, 2000).

[118] Collins, 71-75.

Arno Penzias, who co-directed the project that supplied strong evidence for the Big Bang theory, said, "The best data we have are exactly what I would have predicted, had I nothing to go on but the five Books of Moses, the Psalms, and the Bible as a whole."[120]

Not all secular astronomers were happy with the Big Bang Theory. In *God and the Astronomers*, astrophysicist Robert Jastrow wrote the following:

> *At this moment it seems as though science will never be able to raise the curtain on the mystery of creation. For the scientist who has lived by his faith in the power of reason, the story ends like a bad dream. He has scaled the mountains of ignorance, he is about to conquer the highest peak; as he pulls himself over the final rock, he is greeted by a band of theologians who have been sitting there for centuries.*[121]

Stephen Hawking writes about how the Catholic church was quick to accept the Big Bang Theory because of its God-friendly implications. Hawking himself has been exploring alternatives that he considers less amenable to the idea of a Creator. He wrote that he felt a little uncomfortable presenting a paper on his work to a Jesuit-sponsored scientific conference held at the Vatican.[122] Some readers might find it ironic that Hawking was concerned that he might offend Christians by

---

[119] J.D. Barrow and F.J. Tipler, *The Anthropic Cosmological Principle* (New York: Oxford University Press, 1986), 318. (Also cited by Collins.)

[120] Browne, "Clues to the Universe's Origin Expected," *New York Times*, March 12, 1978.

[121] Robert Jastrow, *God and the Astronomers* (New York: W. W. Norton, 1992), 107 (Cited by Collins)

[122] Stephen S. Hawking, *A Brief History of Time* (New York: Bantam Press, 1998), 121.

speaking against the Big Bang. They might also be surprised that Georges Lematre, one of the first astronomers to suggest the theory, was also a priest.[123]

As with many scientific theories, Big Bang cosmology is under constant scrutiny by those exploring alternative explanations for the evidence. For now, we'll accept it and move on. Following the Big Bang, the scientific creation story continued as matter coalesced into stars and other celestial bodies. The original universe, according to current scientific theory, was composed mostly of hydrogen. One would not have found any rocky planets or heavier elements like iron in the early universe. The fusion processes that took place in the hearts of supernovas are believed to have been the source of heavy elements like oxygen and carbon. As first-generation stars died, these heavy elements were released into the universe. The earth and sun, then, would not have come into being until much later, because heavy elements had not yet been formed. The carbon in the bodies of human beings, according to scientists, was formed inside of ancient stars.[124]

Earth's solar system, according to the current model, formed from a solar nebula, a disk-shaped mass of dust and gas left over from the sun's formation. Astronomers have seen such disk-shaped formations, called protoplanetary disks, around distant stars and believe they are seeing the formation of solar systems.[125] About 4.54 billion years ago, the planet was a

---

[123] Martin Rees. *Universe: The Definitive Visual Guide*. (New York: DK Publishing, 2008), 96, and Adam Hart-Davis. *Science: The Definitive Visual Guide*. (London: Dorling Kindersley, 2009), 320.

[124] Rees, *Universe*, 68.

[125] For information about Hubble images of these disks, see C. Beichman, D. Ardila, and J. Crist. "Spitzer and Hubble Capture Evolving Planetary Systems," posted December 9, 2004, http://hubblesite.org/ newscenter /archive/ releases/ star/protoplanetary-disk/2004/33/ [Accessed June 27, 2012].

molten mass. The surface cooled, and outgassing (gas emerging from deep inside the planet) and volcanic activity produced the planet's primordial atmosphere. Liquid water from larger protoplanets, asteroids, and comets augmented the water that was already present to produce the planet's oceans.

Scientists consider Earth to be a remarkably lucky world in many ways. The planet formed around a single, stable sun. This sun was located far enough from the black hole at the center of the galaxy to be spared the dangers posed by the tidal forces and the debris (Imagine what it would be like for your planet to form near a giant whirlpool in space.). Also, the matter surrounding the sun was not as sparse as it would have been at the outer edges of the galactic disk. The planet Earth, when it formed, happened to be at just the right distance from its sun to receive its warmth without the threat of overheating or deadly bursts of radiation from solar flares. The sun itself was stable and of the right size and type. The immense gravity of large outer planets like Jupiter shielded the growing world from space debris by drawing it away. The planet's rotating core and sliding tectonic plates generated a magnetic field that helped to shield the planet from deadly cosmic radiation. The planet's tilt, stabilized by its large orbiting moon, gave it seasons. The moon's gravity also stirred the ocean tides to prevent the seas from becoming too salty to support life.[126] Some scientists view this convergence of circumstances as evidence of God's providential care, whereas others believe Earth simply happened to win a kind of cosmic lottery.[127]

---

[126] Peter D. Ward and Donald Brownlee, *Rare Earth: Why Complex Life is Uncommon in the Universe* (New York: Copernicus, 2000), xxvii, 1-54.

[127] Lee Strobel, *The Case for a Creator* (Grand Rapids: Zondervan, 2004), 153-192.

CHAPTER 9:
A Guided Tour of the Scientific Timeline, Part II: The Origins of Life, Humanity, and Civilization

The exact details of the emergence of life on earth are unknown. One theory (the *panspermia* theory) suggests that it arrived from space; other theories hold that it developed on earth itself. Some scientists see the immense improbability of life developing on its own as evidence of a guiding intelligence, and others refuse to consider this possibility.[128] Though the idea that life developed spontaneously without the help of a guiding intelligence is often considered a part of evolutionary theory, *abiogenesis*, the emergence of life from nonliving elements, is actually a separate area of study. (Abiogenesis is sometimes referred to as chemical evolution.) Evolution is the process by which life changes and develops over time, but for evolution to take place, life must already exist and have the ability to reproduce and pass along genetic information. These are complex processes even for the simplest of living creatures.

Whatever the actual details of their origin, the earliest living organisms would have found the ancient earth to be a very inhospitable place. With little oxygen and no ozone layer to filter out harmful radiation, the early surface of the planet would have been a hellish environment. The first organisms would likely have appeared in geothermal vents (the fountains of the deep?) that spewed water and steam from beneath the

---

[128] Francis S. Collins, *The Language of God* (New York: Free Press, 2006), 90, 91.

surface.[129] In the Ediacaran Period, multicellular lifeforms with muscular and nervous systems began to appear. Organisms with hard body parts, skeletons, and exoskeletons appeared between the Ediacaran and Cambrian eras. It was during the Cambrian Period, between 543 and 490 million years ago, that many new species appeared suddenly. Vertebrates, including some fish, appeared for the first time.[130] Some Christian scientists see this apparently sudden appearance of so many new lifeforms as evidence of divine action. Francis Collins, a Christian who is also an evolutionist, cautions other Christians against rushing too quickly to the conclusion that God must have intervened miraculously every time scientists encounter something current theory cannot explain. This has sometimes been called the "God of the gaps" approach.[131] Warnings aside, it is difficult for believers in God to look at this sudden emergence of life in the fossil record and not be tempted to view it as God's handiwork in some sense.

Some Christians in the sciences prefer the progressive creation model in which God steps in at key points to miraculously create new species. Others, like Howard Van Till (Not to be confused with Cornelius Van Til, mentioned previously), believe God "planned ahead" and endowed the universe with the qualities it needed to produce life. Van Till uses the expression *fully-gifted creation* to convey this idea.[132]

The epochs that followed, for some, are more challenging to those who see God in the orderly progression of the universe.

---

[129] James F. Luhr, *Earth: The Definitive Visual Guide* (New York: Dorling Kindersley, 2007), 27.

[130] Ibid., 29-31.

[131] Collins, *Language of God*, 92, 93.

[132] See J.P. Moreland and John Mark Reynolds, *Three View on Creation and Evolution*, (Grand Rapids: Zondervan, 1999) for a dialogue. Robert C. Newman represents progressive creationism and Howard Van Till represents "fully gifted creation."

Many new forms of life emerged during those epochs, but the planet was also rocked by a series of mass extinctions. Why, some skeptics have asked, would God keep creating new life and then destroying most of it? As with many things, it is easier to see God in the end results than in the chaos that leads up to them.

An estimated 400 million years ago, plants first appeared on land. The first land animals followed an estimated 370 million years ago. The Mesozoic or "middle life" period, which began about 250 million years ago, was the age of dinosaurs. Many believe it was brought to an end about 65 million years ago, when a 6.2-mile wide meteorite struck ground off the Yucatan Peninsula.[133] The large reptiles were destroyed, but small mammals the size of shrews survived. The Paleocene Epoch witnessed the emergence of large mammals, and in the Eocene, some mammals returned to the ocean, where Basilosaurus is believed to have become the ancestor of dolphins and baleen whales.

Six million years ago, on the scientific timeline, the family lines of humans and chimps diverged from their last common ancestor. The genus homo, from which humans eventually sprung, emerged around 2 million years ago. Homo erectus, a bipedal primate with an average cranial capacity of about two-thirds that of a modern human, began to use fire about 790,000 years ago. Some scientists believe homo habilus or one of the earlier primates may have used it as far back as 1.5 million years ago. Scientists are uncertain whether homo erectus, or even early homo sapiens, had the ability to speak. Neanderthals arrived on the anthropological scene approximately 350,000 years ago[134], shared the planet with modern humans for over two hundred thousand years, and vanished for reasons scientists have yet to agree upon.[135]

---

[133] Luhr, 31-32, and Collins, 95.

[134] Adam Hart-Davis, *History: The Definitive Visual Guide* (New York: DK, 2010), 19.

Neanderthals were the first proto-humans to show evidence of spirituality. Excavations of their ancient graves show that they buried their dead with flowers[136], food, and tools. Though scientists once considered Neanderthals to be the forebears of modern humans, DNA evidence indicates that the two species (or subspecies) were distinct. Neanderthals are no longer believed to be the parents of modern humans but their mysterious older siblings who sprang from a common genetic pool centuries earlier. Though the scientists of earlier eras saw the Neanderthal's stocky build, wide nose, and brow ridge as indications that they were more apelike than modern humans, the size of the Neanderthal brain and studies of Neanderthal relics do not support the "ape-man" image.

Some scientists consider Neanderthals and modern humans to be separate species, referring to them, respectively, as *homo neanderthalis* and *homo sapiens*. Others consider them to be offshoots of the same species, referring to them as *homo sapiens neanderthalis* and *homo sapiens sapiens*. The recent mapping of the Neanderthal genome and its comparison to that of modern humans led to the remarkable conclusion that modern humans of European descent still have traces of Neanderthal DNA in their genetic structure. At some point in the distant past, apparently, interbreeding between the two groups did take place. This would seem to support the belief that modern humans and Neanderthals were, in fact, variants of the same species rather than different species altogether. They would have differed more than the various modern human groups do but would still have been very similar genetically to be capable of interbreeding.

---

[135] Stephen S. Hall, "Last of the Neanderthals," *National Geographic*, October 2008, 36-59.

[136] Arlette Leroi-Goourhan, "The Flowers Found With Shandria IV, A Neanderthal Burial in Iraq," *Science*, Vol. 190, Issue 4214, 562-564.

Anatomically modern humans first appeared on earth, according to current estimates, around 195,000 years ago in Africa.[137] This estimate is based on several remarkable discoveries. One of these came from the study of mitochondrial DNA. Using DNA drawn from people in various parts of the world, scientists found that all living humans shared a single female ancestor. Based on rates and patterns of variation, they placed her in sub-Saharan Africa and estimated that she lived between 100,000 and 200,000 years ago. They named this ancient woman Eve in honor of the biblical Eve.[138] The most recent male ancestor shared by all humans, according to DNA studies, lived in Africa about 50,000 years ago. Scientists have named him Adam. Hugh Ross, of *Reasons to Believe*, suggests that he should have been named Noah instead, since Noah was the father of the nations listed in Genesis.[139]

The oldest fully modern human remains, found in Huerto, Ethiopia, are estimated to be about 160,000 years old.[140] Africa is considered to be the epicenter of the "explosion of humanity" because of the genetic variety of the people groups found there. Scientists believe that the people living in areas of the world that have been inhabited for longer periods of time exhibit a greater variety of genetic differences, whereas areas that have not been inhabited as long will show relatively less variety.

In studying skull morphology and, more recently, DNA, scientists have found the greatest amount of difference among the people groups of Africa, suggesting that their ancestors

---

[137] Collins, *Language of God*, 96.

[138] Dick Fischer, "In Search of the Historical Adam: Part One," *Perspectives on Science and the Christian Faith*, 45 (December 1993) 242.

[139] Hugh Ross, *The Genesis Question* (Colorado Springs: Navpress, 2001), 112.

[140] Hart-Davis, *History*, 19.

went their separate ways many centuries earlier than other groups like the Native American tribes found in North and South America. This is the origin of the Out of Africa theories of human descent. *National Geographic* has undertaken a massive study of human DNA to map the spread of humanity around the globe. This project, called the Genographic Project, yields its own "Table of Nations."[141] Though it supports the idea of humans sharing a common ancestry, it places the dispersion of humans over the face of the earth tens of thousands of years before the emergence of civilization in Mesopotamia.

Fossil evidence, similarly, shows a lengthy span of time between the appearance of anatomically modern human skeletons and the first signs of civilization. Although supposed human ancestors like homo habilus are thought to have used simple tools and made fires because of artifacts found near their bones, Cro-Magnon cave paintings did not appear until around 32,000 years ago. Scientists do not know when humans first developed the ability to speak and use language. One theory links it to the development of a gene. This gene, interestingly enough, has also been found in Neanderthal DNA, so it is possible that they were also capable of speech.[142]

Australian aborigines, according to modern estimates, migrated to the island of Australia 30,000 to 40,000 years ago. At roughly the same time, about 30,000 years ago, Neanderthals vanished from the scene.[143] The cause of their disappearance remains a mystery. Some scientists believe modern humans exterminated them, and others blame a declining birth rate. According to modern estimates, the population of Neanderthals never exceeded more than about

---

[141] Visit the website for the Genographic Project at: https://genographic.nationalgeographic.com/genograpic/ index.html

[142] Hall, *National Geographic*, 5.

[143] Collins, *Language of God*, 96.

13,000 individuals at any given time. This small population alone would have left them vulnerable to extinction from any number of causes.[144] Though Neanderthals are no longer considered to be a bridge between humans and earlier species, they leave us with the same kinds of questions we would ask if we ever encountered intelligent beings—people—on other planets: What was their relationship to God, and is humanity really as special as we once thought? After the disappearance of Neanderthals, modern humans continued to migrate to various parts of the earth. The time of the migration of Native American ancestors into North and South America is estimated to be between 10,000 and 13,000 years ago[145].

According to scientific studies, early humans were beset by a variety of natural disasters. Though geologists find no evidence for a global deluge capable of covering the planet's tallest mountains, early humans were beset by a number of floods and other disasters that threatened their survival. The last Ice Age posed significant challenges for human beings alive during that period of time, and the end of the Ice Age posed a new set of challenges as melting ice led to periodic rises in the levels of the earth's oceans. (So current science supports the idea of planetary flooding, but not a flood that covered all land everywhere simultaneously.) The formation of the Black Sea around 8,000 B.C. flooded thousands of square miles of land in modern Turkey. Author Ian Wilson links the Black Sea flood to the biblical story of Noah.[146] Other authors

---

[144] Adrian W. Briggs, Jeffrey M. Good, Richard E. Green, Johannes Krause, Tomislav Maricic, Udo Stenzel, Carles Lalueza-Fox, Pavao Rudan, Dejana Brajkovik, Zeljko Kucan, Ivan Gusic, Ralf Schmitz, Vladimir B. Doronichev, Liubov V. Golovanova, Marco de la Rasilla, Javier Fortea, Antonio Rosas, and Svante Pääbo[1] "Targeted Retrieval and Analysis of Five Neandertal mtDNA Genomes," *Science,* 17 July 2009: Vol. 325. no. 5938, pp. 318-321.

[145] Hart-Davis, *History*, 29.

[146] Ian Wilson, *Before the Flood* (New York: St. Martin's Press, 2001).

like Carol Hill, however, disagree, placing the biblical flood closer to the development of civilization in the Near East. Geologists once accepted the global nature of Noah's flood as an established scientific fact. With massive amounts of erosion and some fossil beds in various parts of the world, they felt they were on fairly solid ground. In the 1800s, however, new discoveries in the field of geology led to the abandonment of the global flood theory, and many Christians felt that science had betrayed them. (See *The Biblical Flood* by Davis Young for a detailed historical study.[147])

Historians consider Mesopotamia, an area familiar to Bible scholars, to be the "Cradle of Civilization." One theory about the birth of civilization held that agriculture led to permanent settlements, then permanent settlements led to the development of civilization, including institutionalized religion. According to some, the recent discovery of Gobekli Tepe (pronounced Guh-behk-LEE TEH-peh), an amazing structure archaeologists believe ancient people built over a period of centuries between 9600 and 8200 B.C., has threatened to overturn that conclusion. Located south of the mountains from which the Tigris and Euphrates Rivers emerge (one of the proposed locations for Eden, interestingly enough), Gobekli Tepe has been touted by *National Geographic* as the world's first temple. The structure is made up of huge, cleanly-carved limestone pillars decorated with bas-reliefs of animals, including scorpions, foxes, gazelles, boars, and—interesting from a biblical perspective—snakes. The largest of the pillars are eighteen feet tall and weigh about sixteen tons. The structure does not appear to have been located in the middle of a city. The nearest water source is three miles away. It might have been located at the site of a supposed spiritual encounter,

---

[147] Davis A. Young, The Biblical Flood: A Case Study of the Church's Response to Extrabiblical Evidence (Grand Rapids: Eerdmans Publishing Company, 1995).

but without a written record we can only speculate. Archaeologists studying the site wonder if, contrary to the theories that dominated anthropology in the second half of the twentieth century, religion led to permanent settlements, then permanent settlements led to agriculture, not the other way around.[148]

It is too early to say for sure how the discovery of Gobekli Tepe will reshape theories about the development of human civilization. Plant and animal husbandry seem to have developed in that area between 8500 and 7000 B.C. Sumer, one of the great city-states, dominated southern Mesopotamia from 3500-2371 B.C. Sumerians developed complex social structures, invented writing, and used canals to divert water from the rivers to cities located on plains. [149] Ziggurats are found in this area as well. Ur, which many believe to be the biblical home of Abraham, was located just above the Persian Gulf.[150] (There are two Urs in the region, and scholars are divided on which one Abraham came from. The other is in modern Turkey, near the site of Gobekli Tepe.[151]) This area is part of the "Fertile Crescent," an area that runs from northern Africa through Palestine, into southern Turkey, and down the Euphrates River to the Persian Gulf. Great early civilizations developed in this area of the world. Besides the city-states along the Euphrates, other early civilizations appeared in ancient Egypt, in the Indus River Valley (in India), and in China.[152]

---

[148] Charles Mann, "The Birth of Religion," *National Geographic*, Vol. 29(6) (June 2011), 24-59.

[149] Thomas V. Brisco, *The Holman Bible Atlas* (Nashville: Broadman and Holman, 1998), 35-37.

[150] Hart-Davis, *History*, 72-73.

[151] Mann, "The Birth of Religion," 40.

[152] Hart-Davis, *History*, 36-37.

Compared to the relative simplicity of the biblical account of Creation, the scientific account is complex and often confusing. Whereas the Bible is concerned with God as the ultimate cause of everything, science focuses only on the evidence the instruments of science can detect. There may be intermediate physical causes, but Christians will always assume the ultimate cause is spiritual. Scientists, on the other hand, will always seek causes that their science can explain, as long as those causes may be found. That is what scientists do.

Though scientists offer detailed explanations for many phenomena the people of previous generations could only wonder about, there are three great mysteries that remain: the origin of the universe, the origin of life, and the origin of that "divine spark" of consciousness that makes human beings.

CHAPTER 10:
Adam, Evolution, and the Question of Biblical Literary Forms

In studying these two depictions of the development of the universe and the emergence of humanity, we are struck by some remarkable similarities as well as some seemingly problematic differences. As I have noted, people are sharply divided in their reactions to the two narratives. Some Christians and intellectual skeptics alike view them as irreconcilable and mutually hostile.

The recent questions surrounding the "historical Adam" are but one example of the controversies that have divided evangelical leaders. I placed "historical" in quotation marks because Adam was, for all practical purposes, prehistoric. Since Adam's name comes from a Hebrew word, his real name would probably not have been Adam since his life would have predated the Hebrew language. Regardless of one's dating scheme, the biblical account of his life would have been written down many years after his death. It would be more proper to write about the anthropological Adam or the archaeological Adam. Terminology aside, the question is whether Adam was a real man who also serves as a theological symbol or whether he is a theological symbol only. For many evangelicals, the existence of a literal Adam is central to the story of humanity's redemption. For those who accept the old earth timeline, however, finding a place for Adam in that timeline is difficult.

If Adam was the first anatomically modern human male, he would have lived over 100,000 years ago. If he was the first man with modern brain development, he might have lived closer to the era of Cro-Magnon cave paintings. The stories of his children, however, more closely parallel the history of Neolithic Mesopotamia 6,000-10,000 years ago. In the biblical account, Cain is a farmer and Abel a shepherd. Cain's descendents are said to be the fathers of those who dwell in tents, forge metal, and make musical instruments. Agriculture, animal husbandry, and the forging of metal are all said to have begun in Mesopotamia, just as they did in the biblical narrative. These advances spread around the globe from that area. Many of the agricultural plants and animals used worldwide originated in this region of the world.

The idea of a Neolithic Adam is attractive in many ways because it lines up so well with the time and place suggested by the Bible, but it also presents problems. If we are to believe archaeologists, the forebears of the Native Americans and the Australian Aborigines had migrated to their respective locations many centuries before the emergence of civilization in Mesopotamia. Therefore, a Neolithic Adam would not have been the forebear of all of humanity. This is not a problem for Dick Fischer who advocates the Neolithic Adam theory. Fischer proposes that Adam was miraculously created and that his lineage was inserted into the more ancient line of existing humans. This, he says, would explain why other people suddenly show up following Cain's murder of his brother Abel and could shed light on the identity of the mysterious Nephalim. Fischer says:

> *According to the Bible, Adam was the first to have a covenant relationship with the Creator, the first to be accountable, the first to sin and suffer the consequences, and the first in the line of promise leading to the Savior. That does not necessarily mean, however, that Adam was the first biped with an opposable thumb and a cranial capacity of 1300 to 1400 cubic centimeters.*[153]

---

[153] Dick Fischer, "The Search for the Historical Adam, Part 1," *Perspectives on Science and the Christian Faith,* 45 (December 1993): 241.

Naturally, this theory is unacceptable to many evangelicals because it only makes Adam the spiritual or archtypical father of humanity and not its biological father.

Astronomer and apologist Hugh Ross has attempted to develop a moderating view by pushing the emergence of anatomically modern human beings as far forward as the fossil record will allow and allowing for the development of civilization in Mesopotamia at a much earlier date (closer to the time of cave paintings than metallurgy) than most archaeologists would accept strictly on the basis of the physical evidence.[154] Perhaps, Ross might suggest, the remains of those earlier civilizations lie buried under tons of silt deposited by Noah's flood, and what archaeologists see today is only the evidence of humanity pulling itself back up the ladder. This is an interesting possibility, but until such earlier civilizations are unearthed, it will only be an appealing theory and not a true solution. The issues surrounding Adam's place in humanity's primeval timeline are unlikely to be resolved anytime soon, but the intensity of the debate surrounding the biblical first man waxes and wanes across the decades.[155]

The "Adam" question appears in a more generic form in the controversy surrounding the origins of humanity. Since the time of Darwin, the question of human origins has been a divisive issue. It seems, on the surface at least, that it would be simpler for Christians if humans simply showed up in the fossil record about six thousand years ago and if human DNA was markedly different from that of other species. An exploration of the data, as the previous discussion has shown, shows a more complicated picture. Paleontologists have discovered the

---

[154] Hugh Ross, *The Genesis Question*, (Colorado Springs: Navpress, 2001), 107-115.

[155] For an excellent resource on the "Adam" discussions, see Matthew Barrett, and Ardel B. Caneday, *Four Views on the Historical Adam* (Grand Rapids: Zondervan, 2013).

skeletons of other bipedal primates in addition to modern humans. Neanderthal, homo erectus, and homo habilus, in particular, show many human characteristics. Though one or two Neanderthal or homo erectus skeletons might easily have been explained as genetic anomalies, hundreds of such skeletons (over 500, in the case of Neanderthal[156]) are more difficult to dismiss.

Before DNA analysis, some theorized that Neanderthals were merely ice age humans whose skeletal abnormalities were to be explained by conditions like rickets, caused by vitamin deficiencies in people who dwelled in cold climates. Homo erectus, likewise, might merely have been a race of small-brained modern humans. As mentioned previously, studies of Neanderthal DNA have shown that Neanderthals were not modern humans as such but a parallel group believed to have branched off from the same family tree at some earlier time and later interbred with them[157]. Homo erectus may not have been a direct forebear, either; it is difficult to say, because paleontologists do not agree on whether certain skeletons do or do not belong to the homo erectus family. As I have been writing this, a group of scientists has made an interesting study at Dmanisi, a site in Georgia (near Russia, not Florida). They discovered the remains of five hominids that apparently lived at the same time but would have looked very different from each other. According to one of the researchers, if one of the skulls had been found in a different location, he would have thought it belonged to a different species altogether.[158]

---

[156] Amir Aczel, *The Jesuit and the Skull* (New York: Riverhead, 2007), 49.

[157] Green et al. "A Draft Sequence of the Neandertal Genome." *Science*, 7 May 2010. Vol. 328(5979), 710-722.

[158] John Noble Wilford. "Skull Fossil Suggests Simpler Human Lineage." *The New York Times*. October 17, 2013. http://www.nytimes.com/2013/10/18/science/fossil-skull-may-rewrite-humans-evolutionary… [Accessed October 18, 2013].

Though homo erectus seems to have been around more recently than homo habilus, an older species, homo habilus skulls actually look somewhat more like those of modern humans.[159] To the frustration of some evolutionists, paleontologists' attempts to fashion a lineage for modern humans based on these ancient fossils have not yielded the kind of orderly stairstep progression textbooks used to show. With groups developing in isolation from each other, humanity's family tree looks more like a family bush with seemingly newer models predating or coexisting with older ones. Evolutionists, however, are confident that they will figure it all out eventually, and their opponents are equally sure that they never will. So far no one has had any luck reconstructing homo erectus DNA, so researchers have had to rely on bone morphology to identify species and, as the discovery in Georgia shows, that is not an exact science.

Complications aside, the genetic and paleontological evidence at this point seems to indicate that humans are of earth, that humans were made from the same genetic materials as other species. One wonders, however, what might have happened if humans had showed up suddenly in the fossil record and if human DNA had been markedly different from that of other animals. Would all scientists have suddenly started believing in God, or would they have proposed that humans came from another planet? Many would probably have opted for the latter.

Practical issues are also worth noting here. If the structure of human cells had been strikingly different from that of the other plants and animals that live on planet Earth, these unique beings would not have been able to survive. The human body is precisely geared to break down certain types of chemicals. The complexity of this process is never more obvious than when it

---

[159] *Encyclopædia Britannica Online*, s. v. "Homo erectus," http://www.britannica.com/ EBchecked/topic/270386/Homo-erectus. [accessed June 12, 2012].

doesn't work properly. Conditions like diabetes, lactose intolerance, and gluten intolerance are painful examples of what happens when this fine-tuning fails.

For humans to survive and thrive, we would have to be able to digest the stuff of earth. To do this, we would have to share many of the biological properties seen in the species of earth. The question is whether God designed humans suddenly with these features in place or whether humans developed gradually by a natural process that could have been ordained by God. Some Christians would argue that it doesn't really matter, but others would fight as though the very truth of Scripture depended upon humanity's spontaneous creation. The fear is that interpreting the early Genesis narratives figuratively ultimately leads to one interpreting all of the Bible's miracle stories figuratively, and some have done that.

Before the discussion continues, however, it must be pointed out that the Bible never insists that humans are not a part of the same creation as other animal species, but that they hold a special place in that creation. Although the story of Adam's creation might give the impression of human beings springing up apart from the rest of nature, the creation week narrative places them at the apex of God's ongoing creative process. They are of God, but they are also of earth in the sense that they arose as a part of the same process by which God created the rest of the species on earth. They are the dust of earth infused with the breath of God.

If humans did not evolve from more primitive species, one might ask, why are there skeletons that resemble us? How do australopithecus, homo habilis, homo ergaster, homo erectus, and the rest fit into the Christian worldview? Were there, at the same general time, varieties of apes that walked upright and varieties of humans who had smaller brains? Was the similarity a matter of superficial morphology and adaptation to the same environmental conditions, rather than ancestry? This is a question better answered by geneticists as genetic data continues to become available.

Could the Christian view of humanity survive if a genetic link were proven, beyond all reasonable doubt, to exist? (I know. Some say it already has, and some laugh at the very idea.) For theologians like B.B. Warfield, the answer is a resounding yes:

> *We may be sure that the old faith will be able not merely to live with, but to assimilate to itself all facts.... The only living question with regard to the doctrine of evolution still is whether or not it is true.*[160]

Others are not so sure. Some Christian organizations teach that anyone who believes in evolution is not a true Christian. A brief online search for opinions on the subject shows that Christians now, as in 1895 when Warfield published his declaration, are deeply divided over the issue of evolution.

Controversies aside, not all Christians view faith and science as opposing forces. At least, they do not have to be. Many Christians view the two timelines as largely complementary pictures of the universe: The biblical picture is sacred and timeless. It provides a theological portrait of a loving and transcendent God and his relationship to humanity. The scientific picture is a work in progress. It represents a point in humankind's unending journey to understand the marvels and mysteries of a God-ordained universe. Those who hold this view believe it is possible to get at least a partial understanding of origins through clues left behind. The heavens serve as a window to the past, and the rocks "cry out" with data to be studied. To understanding the meaning of it all, however, one must look to God's revelation in the Bible.

Canadian astronomer Hugh Ross, as a young skeptic, read the creation stories of a variety of religions and was not impressed by them. Later, as a Christian, he remembered reading through Genesis for the first time:

---

[160] Benjamin Warfield, "The Present Status of the Doctrine of Evolution." *The Presbyterian Message.* 3(10), December 5, 1895, pp. 7-8.

> *The book's distinctives struck me immediately. It was simple, direct, and specific. I was amazed with the historical and scientific references and with the detail in them... Instead of another bizarre creation myth, here was a journal-like record of the Earth's initial conditions —correctly described from the standpoint of astrophysics and geophysics... The account was simple, elegant, and scientifically accurate. From what I understood to be the stated viewpoint of an observer on Earth's surface, both the order and description of creation events perfectly matched the established record of nature. I was amazed.*[161]

The only real discrepancies Ross saw between the modern scientific model and the Bible's ancient narrative were the spans of time involved and the late creation of the sun, moon, and stars. Ross was not really bothered by either of these details. *Yom*, the Hebrew word for day, has some of the same flexibility as its English counterpart.[162] One might use it to refer to a twenty-four hour period or an era, as in "the day of the steam engine." Ross understood the passage referring to the late appearance of the sun, moon, and stars was to be read as though it were written by someone standing on earth, watching as a translucent early atmosphere became transparent and celestial bodies appeared for the first time.

Ross's testimony is encouraging to Christians who work in scientific fields, but some Bible scholars question whether his approach is really the best way to interpret the narrative. Ross is viewing the creation story through the eyes of a modern scientist, and it actually works fairly well in the case of the

---

[161] Hugh Ross, *The Creator and the Cosmos*" (Colorado Springs: Navpress, 1993), 19.

[162] Hugh Ross, *The Genesis Question* (Colorado Springs: Navpress, 2001), 65.

creation week narrative (though the "evening and morning" statements still have to be taken metaphorically.) When he tries to explain the creation of Eve from a rib (or side. The Hebrew word *tsela* can be translated either way.) in scientific terms and explains that the rib was really more like a biopsy,[163] the framework does not seem to fit as comfortably. He does say, however, that God's choice to fashion Eve from tissue taken from Adam was meant to communicate a message about the relationship between genders. He considers it a literal act with a symbolic message.[164]

Ross's approach is an example of an approach called *concordism*. This approach is based on the assumption that Genesis was trying to communicate literal, scientific facts but expressing them in the language of the day. Unconsciously, he is still assuming that the original readers were looking for a scientific explanation of reality. As we have seen, however, the biggest threat to belief in God at the time Genesis was written was not scientific atheism but paganism. For Genesis to catch on in an age dominated not by scientific theories but by epic myths, an unadorned recitation of data was not enough. The story had to have some beauty and drama.

Some have observed that the narrative style of Genesis is typical of that found in oral cultures. People in cultures that do not use written language as the primary means of communication, according to Hiebert, "store information in songs, poems, proverbs, riddles, chants, and stories, all of which aid recall."[165] Contrast, if you will, the styles of the creation week narrative in Genesis to Paul's epistles. The creation narrative uses repeated phrases and simple imagery

---

[163] Ibid., 75.

[164] Ibid.

[165] Paul G. Hiebert, *Antropological Insights for Missionaries*. (Grand Rapids: Baker Book House, 1985), 161.

that make it easy to remember and visualize. This is the type of story one would use to communicate doctrine to an oral culture. Paul's letters were "born" as written documents. His sentences are longer and more complicated, and he uses fewer word pictures and expresses spiritual concepts in more abstract ways. They would have been harder to memorize, but they did not have to be since they were available in written form.

Eugene Merrill, in the (Southern Baptist) *Holman Concise Bible Commentary* expresses the same idea when he describes the first chapters of Genesis in this way: "The primeval events are cast in a poetic narrative form to aid in oral transmission."[166] He adds, "Though the creation stories are fundamentally theological and not scientific, nothing in them is contradicted by modern scientific understanding."[167]

Most modern westerners don't really stop and consider the kinds of issues that would have concerned the original readers of Genesis. We understand the rotation of planets, the moon cycle, and weather patterns. As far as the ancient people were concerned, all of these things were purely at the pleasure of the gods. They got nervous when the moon went away at the end of the month, because they were afraid it might not come back. The Sabbath, based on the lunar cycle, was a serious time for them. If they didn't please the gods, they might lose the moon. Eclipses terrified them, because the sun was not supposed to be dark in the middle of the day. What if the gods decided to take away the sun? Rainfall was also at the pleasure of the gods. Rainfall might stop forever, or it might keep falling until it flooded the entire world out of existence. The idea of God establishing boundaries for the waters and ordering the days, nights, and seasons is something we take for granted, but it was a serious concern to people back then. To some primitive

---

[166] Eugene H. Merrill, "Genesis," *Holman Concise Bible Commentary*, ed. David S. Dockery (Nashville: Broadman and Holman, 1998), 6, 7.

[167] Merrill, 8.

people-groups, it still is. I once talked to a soldier who had been stationed in Afghanistan about the beliefs of the people there. As hard as it is for us to imagine, he said some of the people in the region where he was stationed thought a dragon swallowed the sun at the end of every day and spit it back out every morning. That was their understanding of the universe.

Some have suggested that the strength of the Bible is not that it is a better science book, but that it is not a science book at all. Scientific theories, after all, have a "shelf life" and not being tied to one particular generation's form of science offers the Genesis epic one major advantage: longevity.

At the time that Genesis was written, people believed in a flat earth with a bowl-shaped sky and a freshwater ocean overhead. The Greek model replaced the flat earth with a spherical earth that sat at the center of the universe. Since Galileo, modern people have believed in a spherical earth orbiting a sun. Newton discovered universal gravitation but believed in a static universe.[168] Darwin taught that biology, contrary to the "Watchmaker" model put forth by Paley, is in a constant state of change. Though there are disagreements on the exact mechanisms of that change and the extent to which it explains the variety of species existing on the earth today, there is no question that plants and animals experience variation over time and in response to the environment. While scientists and theologians were still reacting to Darwin's theories, Einstein taught that even time and space are relative.

Though these changes led to varying levels of theological upheaval, Christianity has managed to survive. The Bible, many of us would argue, has just as much to say now as it did when it was written, not because it teaches better science but because it transcends science. Science is about the details, but the Bible concerns itself more with meaning and purpose. Science is an excellent source of information, but it is not always the best source of wisdom or of truth.

---

[168] Charles E. Hummel, *The Galileo Connection* (Grand Rapids: Intervarsity Press, 1986), 256.

Genesis was originally revealed to "flat earthers." It did nothing to correct their primitive cosmology but was nothing short of revolutionary in its portrayal of a transcendent God who existed before and beyond the created universe. Perhaps, modern people think, if God had *really* revealed Genesis to the people who lived over three thousand years ago, He would have given them a "modern," 21st- century model of the universe, never realizing that our current view of the universe will one day be as outdated as those of our ancestors.

When the current form of Big Bang cosmology is replaced by whatever comes after it and our current theories about human origins have been upgraded to whatever comes after them, chances are that the truths of Genesis will continue to speak to the hearts of those who read its panoramic story of a powerful and loving God.

CHAPTER 11:
Full Circle

So where does all of this leave us, and where do we go from here? Let's take a quick review of our journey. In my introduction, I mentioned three ways Christians have viewed the relationship between faith and reason.

The first, *reason then faith*, assumes that Christians arrive at their beliefs rationally and intellectually after weighing the various alternatives.

The second, *faith alone*, assumes that faith decisions are primarily "heart" decisions and that reason does not and should not play any kind of major role. Calling it "faith alone" is really misleading because nobody really believes anything for no reason. All beliefs are based on some kind of evidence, but "faith alone" is not preceded by the kind of deliberate, intellectual analysis that we see with the first approach.

The third position, *faith seeking understanding*, is an attempt to integrate the two. It recognizes that most people do not arrive at their beliefs through the kind of logical process described by the "reason then faith" scenario, but that "faith alone" is often insufficient if one's faith is to survive in a world filled with conflicting truth-claims. One can either withdraw from the world into an isolated faith community or try to develop some intellectual defenses.

As I pointed out earlier, not many people reason their way to faith, but too many unanswered questions can ultimately lead to a loss of belief or, at least, a loss of heart. Christians who work in fields relating to biology, archaeology, geology, and astronomy encounter intellectual challenges to their faith on a regular (perhaps daily) basis, but often receive little guidance in dealing with them. Scientists and other academic/intellectual types are a relatively small percentage of the population, and few churches really know what to do with them. This is both unfortunate and short-sighted because they are also a highly influential group. University professors shape the minds of entire generations of professional people, and they shape society as a whole. The failure of evangelicals to pass along their intellectual heritage (Yes, we do have one, believe it or not.) to a new generation of scholars is a recipe for cultural suicide.

Recent movements have called for Christians to seek the "heart of worship" and to express their faith through ministry. I applaud these movements, but I also want to encourage the development of a new spirit of intellectual engagement among evangelicals. That goal is at the heart of what I hoped to accomplish in writing this book.

At the beginning of my first chapter, I wrote about the clashes over textbooks that have taken place in the legislatures across the United States and about the movement toward homeschooling. One reason, I believe, that those efforts have not been effective in "bringing God back into science" is a failure to consider background issues and assumptions. These include such practical issues as the make-up of the scientific community. The communities of scientists that decide the direction of science (e.g. the Planetary Society ruling on issues like whether or not Pluto should be considered a planet) operate in mostly secular organizations. Science, for them, is a faith-neutral enterprise. Political efforts to introduce religious views into these organizations are usually doomed at the outset.

There are also differences in goals. If Christians are eager to give God credit for miraculously creating the universe, and scientists want to learn how things work, their goals are incompatible. Christians want to play "stump the scientist," and scientists refuse to admit defeat. Christians in scientific fields learn to live with the tension of believing in the possibility of miracles but looking for natural explanations and seeing God at work in the natural as well as the unexplainable.

I also wrote about the questions surrounding the nature of human reason and the nature of reality. It is popular in some Christian circles to teach that human reason was so badly ruined by humanity's fall that it is completely unreliable. When reason conflicts with the Bible, the Bible must prevail. This is often taken to mean that when reason conflicts with a literal reading of the Bible, the literal reading of the Bible must prevail. I understand the reason for this. There are some scholars who view all of the Bible's miracle stories as metaphors because they do not believe in the supernatural. I would argue that both of these "one size fits all" approaches short-circuit the process of Bible interpretation. The literary form of the book in question has to be taken into consideration. The gospels and the book of Revelation, for example, use wildly different literary forms.

In faith-versus-science discussions, the "always literal" argument is sometimes used to defend wildly implausible theories and a generally "fast and loose" attitude toward science. (This is weighed against taking a "fast and loose" attitude toward Bible interpretation which is a serious concern for Christians, of course.) I argued that those who have such a low opinion of science would be better off avoiding it entirely rather than pretending to offer scientific defenses of the Bible. Though Christian laypeople often respond positively to groups that demonize scientists, those groups also serve to push Christian scientists and academics out the door of the church and breed new generations of angry skeptics.

I also mentioned questions about the nature of reality itself. One Sunday school teacher I know used to tell high school students that God put dinosaur bones into rocks to test the faith of Christians. This is a street example of the "appearance of age" theory that some Christians use to resolve the apparent inconsistencies between the genealogies given in the Bible and the geological and astronomical evidence for an ancient earth. As a physicist friend of mine pointed out, God could have done it that way. He could also have created us all five minutes ago, and given us false memories of lives we never lived. The question is whether it is more of a theological problem for the genealogies to have gaps in them or for God to create false evidence for an ancient earth in a deliberate attempt to deceive unbelievers.

In Chapter 2, I dealt with the most serious problem Christian academics face when they are exposed to new evidence: Religion demands dedication to doctrine and academic honesty requires one to be indifferent about results and follow the evidence wherever it happens to lead. If the Bible is really true, of course, this should not be a problem. A Christian academic should always follow the evidence where it leads and honestly report the results. That may result in some short-term problems, but the faith of Christians is that everything will ultimately work out in the end. Most Christians would agree with this, but they differ in how they expect everything to work out. Strict literalists believe following the evidence will ultimately lead back to proof of a 6,000-year-old earth, a planetary flood, and a six-day creation. Other Christians allow for some symbolic language in the Genesis account, but believe the Bible and nature both come from God and that the apparent conflicts will turn out to be nothing more than differences in literary packaging. The question for them is how far one can adapt one's understanding of the Bible to accommodate new discoveries without sacrificing its authority. Obviously such adaptations have taken place over the centuries. The sixth century debate between round earth and flat earth Christians sounds a great deal like modern arguments over the age of the earth, but the church managed to accommodate a round earth without too much harm to the authority of Scripture. The willingness of some liberal Protestants to treat the resurrection of Jesus from the dead as a symbolic rather than literal event, however, makes many evangelicals understandably skittish about treating any event described in the Bible as anything but straight history.

In the chapters that followed, I took you on a tour of two parallel timelines. I led you on a commentary-style guided tour of the creation events described in Genesis first. Then we took a trip through the scientific timeline, starting with the Big Bang and ending with the emergence of civilization. Some readers, no doubt, saw the two accounts as "Christian" and "secular," but many Christian scientists see God in the "secular" picture too. Note, in particular, the comments of the scientists involved in researching Big Bang cosmology. As I have noted previously, the three biggest questions in the scientific timeline concern the origin of the universe, the origin of life, and the origin of human consciousness. Scientists continue to look for scientific explanations to these questions, of course, but Christians ultimately trace the "big three" back to God even if science supplies some of the mechanical details.

So, where does all of this leave us? "If you can't tie all of the problems up into a nice, neat package," some might ask, "why write about the issues at all?" To be honest, I have asked myself the same question, but I received my answer a few weeks ago when I had a conversation about some of these issues with a biology professor I know. I told him about the research I'd been been doing and sent him a paper I'd written. A few hours later, he sent me an email message: "This really helped me in my faith. Now I don't feel like such a contradiction." Given his dry sense of humor, I though he might be joking, but he told me he was completely serious. Even if I had not been able to resolve all of the problems, my exploration of the issues had been helpful to him just the same, and I hope it has been to you too.

Bibliography

Books:

Aczel, Amir. *The Jesuit and the Skull.* New York: Riverhead, 2007.

Alexander, Denis. *Rebuilding the Matrix.* Grand Rapids: Zondervan, 2001.

Barrett, Matthew, and Ardel B. Caneday, *Four Views on the Historical Adam.* Grand Rapids: Zondervan, 2013.

Barrow, J.D. and F.J. Tipler, *The Anthropic Cosmological Principle.* New York: Oxford University Press, 1986.

Barstone, Willis. *The Other Bible.* New York: HarperOne, 2005.

Boa, Kenneth D. and Robert M. Bowman. *Faith Has Its Reasons.* Grand Rapids: InterVarsity Press, 2001.

"The Book of the Secrets of Enoch," quoted in Willis Barnstone, *The Other Bible.* New York: HarperCollins, 2005.

Brand, Chad Owen, and David E. Hankins, *One Sacred Effort.* Nashville: Broadman and Holman, 2005.

Brisco, Thomas V. *Holman Bible Atlas.* Nashville: Holman, 1998.

Collins, Francis S. *The Language of God.* New York: Free Press, 2006.

Criswell, W.A. *The Criswell Study Bible.* Nashville: Zondervan, 1979.

Cuddihy, John Murray. *No Offense: Civil Religion and Protestant Taste*. New York: Seabury, 1978.

Dawkins, Richard. *The Blind Watchmaker.* New York: Norton, 1986.

Dockery, David, General Ed., *The Holman Concise Bible Commentary.* Nashville: Broadman and Holman, 1998.

Elliott, Ralph H. The Message of Genesis: A Theological Interpretation. Nashville: Broadman Press, 1961.

Eydoux, Henri-Paul. "The Men Who Built the Tower of Babel," in *The World's Last Mysteries.* Pleasantville, NY: The Reader's Digest Association, 1978.

Francisco, Clyde T. "Genesis," in *The Broadman Bible Commentary, Vol. 1 General Articles, Genesis, Exodus.* Clifton J. Allen, General Editor. Nashville: Broadman Press, 1973.

Garrett, Duane. *Rethinking Genesis.* Grand Rapids: Baker, 1991.

Gonzales, Justo L. *The Story of Christianity*, Volume 1. New York: HarperCollins, 2010.

Hamilton, Victor. The Book of Genesis, Chapters 1-17, vol. 1 of The New International Commentary on the Old Testament .Grand Rapids: William B. Eerdmans Publishing Company, 1990.

Hart-Davis, Adam. *History: The Definitive Visual Guide*. New York: DK, 2010.

Hart-Davis, Adam. *Science: The Definitive Visual Guide.* London: Dorling Kindersley, 2009.

Hawking, Stephen S. *A Brief History of Time.* New York: Bantam Press, 1998.

Hiebert, Paul G. *Antropological Insights for Missionaries.* Grand Rapids: Baker Book House, 1985.

Hummel, Charles E. *The Galileo Connection.* Downer's Grove, IL: Intervarsity Press, 1986.

Hoerth, Alfred J. *Archaeology and the Old Testament.* Grand Rapids: Baker Academic: 1998.

Hunter, James Davison. *Culture Wars.* New York: Basic Books, 1991.

Hyers, Conrad. *The Meaning of Creation.* Atlanta: John Knox Press, 1984.

Jastrow, Robert. *God and the Astronomers.* New York: W. W. Norton, 1992.

Ken Keathley, "Flat or Round? The Sixth Century Debate Over the Shape of the Earth," *Intelligent Design: William A. Dembski and Michael Ruse in Dialogue*, ed. Robert Stewart. Minneapolis: Fortress Press, 2007.

Keller, Werner. *The Bible as History.* New York: Bantam, 1982.

Kinnaman, David. *You Lost Me.* Grand Rapids: BakerBooks, 2011.

Kovacs, Maureen Gallery. *The Epic of Gilgamesh.* Stanford: Stanford University Press, 1989.

Kuhn, Thomas S. *The Structure of Scientific Revolutions.* Chicago: The University of Chicago Press, 1962.

Lasor, William S.; David A. Hubbard, and Frederic W. Bush, *Old Testament Survey.* Grand Rapids: Eerdmans Publishing Co, 1996.

Liddle, Andrew R. and David H. Lyth, *Cosmological Inflation and Large-Scale Structure.* Cambridge: Cambridge University Press, 2000.

Luhr, James F. *Earth: The Definitive Visual Guide* (New York: Dorling Kindersley, 2007.

McGrath, Alister. *Dawkins' God: Genes, Memes, and the Meaning of Life.* Malden, MA: Blackwell Publishing, 2007.

McIntyre, C.T. "The Fundamentals," *Evangelical Dictionary of Theology.* Walter Elwell, ed. Grand Rapids: Baker Academic, 2001.

McKim, Donald K. *Westminster Dictionary of Theological Terms.* Louisville, KY: Westminster John Knox Press, 1995.

Merrill, Eugene H. "Genesis," *Holman Concise Bible Commentary*, ed. David S. Dockery (Nashville: Broadman and Holman, 1998), 6, 7.

Michener, James A. *Hawaii.* New York: Bantam, 1957.

Moore, Randy; Mark Decker, and Sehoya Cotner, *Chronology of the Evolution-Creation Controversy.* Santa Barbara, CA: ABC-CLIO, LLC, 2010.

Moreland, J.P. and John Mark Reynolds. *Three View on Creation and Evolution.* Grand Rapids: Zondervan, 1999.

Noll, M.A. "Warfield, Benjamin Breckenridge," *Evangelical Dictionary of Theology*, 2nd Edition. Walter Elwell, ed. Grand Rapids: Baker Academic, 2001.

Newport, John Newport, *Life's Ultimate Questions*. Dallas: Word Publishing, 1989.

Ramm, Bernard. *The Christian View of Science and Scripture*. Grand Rapids: William B. Eerdmans Publishing Company, 1954.

Rees, Martin. *Universe: The Definitive Visual Guide*. New York: DK Publishing, 2008.

Rennie, I.S. "Verbal Inspiration," *Evangelical Dictionary of Theology*, 2nd Edition. Walter Elwell, Ed. Grand Rapids: Baker Academic, 2001.

Ross, Hugh. *Creation and Time*. Colorado Springs: Navpress, 1994.

Ross, Hugh. *The Creator and the Cosmos*. Colorado Springs: Navpress, 1993.

Ross, Hugh. *The Genesis Question*. Colorado Springs: Navpress, 2001.

Smith, Christian. *Souls in Transition*. Oxford: Oxford University Press, 2009.

Stafford, Adam. *The Adam Quest*. Nashville: Nelson Books, 2013.

Stewart, Robert B. *The Resurrection of Jesus*. Minneapolis, MN: Fortress Press, 2006.

Strobel, Lee. *The Case for a Creator.* Grand Rapids: Zondervan, 2004.

Waltke, Bruce K. *Genesis: A Commentary.* Grand Rapids: Zondervan, 2001.

Ward, Peter D. and Donald Brownlee, *Rare Earth.* New York: Copernicus, 2000.

Wenham, Gordon J. Genesis 1-5, vol. 1 of the *Word Biblical Commentary.* Nashville: Thomas Nelson, 1987.

Whorton, Mark, and Hill Roberts. *Holman QuickSource Guide to Understanding Creation.* Nashville: Broadman and Holman, 2004.

Wilson, Ian. *Before the Flood.* New York: St. Martin's Press, 2001.

Wuthnow, Robert. *The Restructuring of American Religion.* Princeton, NJ: Princeton University Press, 1988), 155. Cited in: James Davison Hunter. *Culture Wars.* New York: HarperCollins, 1991.

Yancey, Philip. *I Was Just Wondering.* Grand Rapids: William B. Eerdmans Publishing, 1989.

Yancey, Philip. *What Good is God?* New York: Faithwords, 2010.

Young, Davis A. The Biblical Flood: A Case Study of the Church's Response to Extrabiblical Evidence. Grand Rapids: Eerdmans Publishing Company, 1995.

Periodicals:

Briggs, Adrian W., Jeffrey M. Good, Richard E. Green, Johannes Krause, Tomislav Maricic, Udo Stenzel, Carles Lalueza-Fox, Pavao Rudan, Dejana Brajkovik, Zeljko Kucan, Ivan Gusic, Ralf Schmitz, Vladimir B. Doronichev, Liubov V. Golovanova, Marco de la Rasilla, Javier Fortea, Antonio Rosas, and Svante Pääbo[1] "Targeted Retrieval and Analysis of Five Neandertal mtDNA Genomes," *Science,* 17 July 2009: Vol. 325. no. 5938, pp. 318-321.

Browne, "Clues to the Universe's Origin Expected," *New York Times*, March 12, 1978.

Hill, Carol A. "The Garden of Eden: A Modern Landscape," *Perspectives on Science and Christian Faith*, 52 (March 2000), 31-46.

Fischer, Dick. "In Search of the Historical Adam: Part One," *Perspectives on Science and the Christian Faith*, 45 (December 1993) 242.

Green, Richard et al. "A Draft Sequence of the Neandertal Genome." *Science*, 7 May 2010. Vol. 328(5979), 710-722.

Hall, Stephen S. "Last of the Neanderthals," *National Geographic*, October 2008, 36-59.

Hill, Carol A. "Making Sense of the Numbers of Genesis," *Perspectives on Science and Christian Faith*. Vol. 44, No. 4, (Dec 2003), pp. 239-251.

Jastrow, Morris. "The Hebrew and Babylonian Accounts of Creation." *The Jewish Quarterly Review* 13 (July 1901): 626-630.

Leroi-Goourhan, Arlette. "The Flowers Found With Shandria IV, A Neanderthal Burial in Iraq," *Science*, Vol. 190, Issue 4214, 562-564.

Mann, Charles. "The Birth of Religion," *National Geographic*, Vol. 29(6) (June 2011), 24-59.

Newcomb, Simon. "The Place of Astronomy Among the Sciences," *The Sidereal Messenger*, 7 (1888): 69-70.

Warfield, Benjamin . "The Present Status of the Doctrine of Evolution." *The Presbyterian Message*, December 5, 1895, 7-8.

Electronic Media:

Rendsberg, Gary A. "Lecture 2: Genesis 1," *The Book of Genesis*. The Great Courses audio lecture series, produced by The Teaching Company of Chantilly, VA. 2006.

Online Sources:

Beichman, C., D. Ardila, and J. Crist. "Spitzer and Hubble Capture Evolving Planetary Systems," posted December 9, 2004, http://hubblesite.org/ newscenter /archive/ releases/ star/protoplanetary-disk/2004/33/ [Accessed June 27, 2012].

Encyclopædia Britannica Online, s. v. "Homo erectus," http://www.britannica.com/ EBchecked/topic/270386/Homo-erectus. [accessed June 12, 2012].

Honey, Charles. "Adamant on Adam," *Christianity Today*, posted: 5/25/10, http://www.christianitytoday.com/2010/june/1.14.html [accessed August 18, 2011].

Lovan, Dylan. "Top Home-School Texts Dismiss Evolution for Creationism." *USA Today.com*. Posted 3/8/2010, 2:30 p.m. http://www.usatoday.com/news/religion/2010-03-08-home-school-christian_N.htm [Accessed July 9, 2012].

Newport, Frank. "In U.S. 46% Hold Creationist View of Origins," Gallup Organization, entry posted June 1, 2012, http://www.gallup.com/poll/155003/Hold-Creationist-View-Human-Origins.aspx [accessed July 9, 2012].

"OT Scholar Bruce Waltke Resigns Following Evolution Comments." *Christianity Today*. Posted on April 9, 2010. http://blog.christianitytoday.com/ctliveblog/archives/2010/04/ot_scholar_bruc.html

Roach, David. "How Old? Age of Earth Debated Among SBC Scholars," *Florida Baptist Witness*, Oct 20, 2010.

Wilford, John Noble. "Skull Fossil Suggests Simpler Human Lineage." *The New York Times*. October 17, 2013. http://www.nytimes.com/2013/10/18/science/fossil-skull-may-rewrite-humans-evolutionary... [Accessed October 18, 2013].

www.ingramcontent.com/pod-product-compliance
Lightning Source LLC
Chambersburg PA
CBHW031448040426
42444CB00007B/1024